Measurement of Dissolved Oxygen

CHEMICAL ANALYSIS

A SERIES OF MONOGRAPHS ON
ANALYTICAL CHEMISTRY AND ITS APPLICATIONS

Editors

P. J. ELVING · J. D. WINEFORDNER

Editor Emeritus: I. M. KOLTHOFF

VOLUME 49

A WILEY-INTERSCIENCE PUBLICATION

JOHN WILEY & SONS

New York / Chichester / Brisbane / Toronto

Measurement of Dissolved Oxygen

MICHAEL L. HITCHMAN

Laboratories RCA Ltd.
Zurich, Switzerland

This book was commissioned by Orbisphere Laboratories, a division of Orbisphere Corporation, and was begun while the author was an employee of the Laboratories. Orbisphere personnel have contributed to it by providing advice, experimental data, and reference material.

A WILEY-INTERSCIENCE PUBLICATION

JOHN WILEY & SONS

New York/Chichester/Brisbane/Toronto

Library of Congress Cataloging in Publication Data

Hitchman, M. L.
  Measurement of dissolved oxygen.

  (Chemical analysis; v. 49)
  "A Wiley-Interscience publication."
  Includes bibliographies and index.
  1. Water—Dissolved oxygen—Measurement.   I. Title.

II. Series
QD142.H57   546'.721'5483   77-26710
ISBN 0-471-03885-7

Printed in the United States of America

10 9 8 7 6 5 4 3 2 1

To three generations—
my parents, my wife, my children

FOREWORD

Dr. Hitchman in his Introduction rightly submits that the importance of a subject, such as the quantitative assay of oxygen, may be related to the quantity of papers it generates over long time periods. On the other hand, some substantial scientific problems are so difficult and perhaps intractable that they entice few workers and as a consequence generate few results. Such a problem is presented in determining both the mechanisms by which the oxygen molecule oxidizes substances in homogenous solutions and the identity of the chemical species involved. And herein will lie the real value of this volume; it will be a stepping stone for some scientists to attack these next generation problems.

The intimate involvement of oxygen in life processes, photosynthesis, and respiration, has induced many scientists to determine its concentration in a variety of environments for scientific or technological studies. In my own field of oceanography, there are perhaps more measurements of oxygen in seawater than of any other chemical except for total salt contents. Marine chemists have been forced to confront a practical compromise between the rapidity in analysis versus the precision of the results or the ease of applicability for a given technique to solutions of varying salinity and composition. More recently, the need for *in situ* measurements has placed new demands on the electrochemical techniques where the proposed instrument may encounter a range of temperatures and pressures in the field.

But the marine chemists, as well as their colleagues in other scientific pursuits, are often bothered and bewildered by their inabilities to explain how the diatom oxygen carries out its chemistries. The redox potentials of seawater are yet to be defined. There is a sense that the O_2/H_2O_2 couple plays a substantial role in the redox chemistry of waters, but the levels of the peroxide or even its existence are yet to be demonstrated definitively to the environmental scientist.

Perhaps the chemists of the next decades will remove the clouds of uncertainty surrounding the behavior of oxygen in natural systems. This volume, which provides the pathways to the measurement of oxygen, with delightful injections of humor, will inspire some workers to take on these difficult problems.

EDWARD D. GOLDBERG

Scripps Institution of Oceanography
La Jolla, California
February 1978

vii

PREFACE

One of the characteristics of our age is the way in which a great deal of time and energy is taken to quantify and measure anything and everything. Sometimes this effort is of doubtful value: for example, the U.S. Federal Aviation Administration has recently issued an exhaustive catalogue of the measurements of more than 450 air hostesses, ranging from the length of the noses to the depths of the buttocks. On other occasions the effort and its result are nothing short of ludicrous: a book entitled *The French Such As They Are*, based on nearly 2000 interviews, reveals that 20% of the French have a barometer in their bedroom, but only 14% have a chamber pot.

One area of measurement, however, which is certainly neither of doubtful importance nor trivial, is the monitoring of dissolved oxygen (DO). Chapter 1 of this book tells us why. There are six major areas that directly or indirectly affect our everyday lives and in which it is necessary to know the level of DO: environmental control of natural waters, sewage waste treatment, medical and biochemical studies, soil husbandry, food and drug process control, and prevention of corrosion in boilers. The need and importance of DO measurements in these areas constitute the raison d'être for this book.

Now it could be argued that if these specific situations are so important then in subsequent chapters more space than I have allowed should have been devoted to describing and considering the special techniques and applications of DO measurements in such situations. However, in preparing this book I chose not to take this course for two main reasons. First, to have written an extensive review of DO measurements in all of the areas mentioned would have been a mammoth task, and having had experience in a limited area of application only—environmental monitoring—I did not think I was especially qualified to undertake this. Second, in the last few years a number of useful reviews and proceedings of symposia have been published which present details of the methodology and applications of measurements of oxygen, and to have given a comprehensive coverage here of monitoring techniques would have been, it seems to me, repetitive and superfluous. Anyone who is interested in particular techniques should consult the review publications referenced at the end of Chapters 5, 7, and 8. Especially recommended are the proceedings of the Odense Interdisciplinary Symposium (H. Degn, I. Balslev, and R. Brook, Eds., Elsevier Scientific Publishing Company, 1976), the Oxygen Pressure Recording Symposium in

Nijmegen (F. Kreuzer, Ed., S. Karger, 1969), and the workshop on Oxygen Supply in Dortmund (M. Kessler et al., Eds., Urban & Schwarzenberg, 1973). Also Fatt's detailed treatment of the clinical applications of polarographic oxygen sensors (I. Fatt, CRC Press, 1976), which gives a valuable insight into the nature of the problems encountered and the solutions found in one particular area, should prove to be a very useful source of practical information for a long time to come. What I have tried to do in this book, however, is to assess critically the various aspects of DO measurements and, more particularly, to emphasize the physicochemical principles underlying such measurements. These principles are generally applicable and remain the same whether one is monitoring DO in a tissue, in sewage, in fruit juice, or wherever. Technique and methodology are, of course, always necessary and important, but they become more powerful and productive if the principles behind them are clearly understood. This book therefore tries to deal with the "why" of DO measurements rather than the "how."

Thus in Chapter 2, for example, I discuss the thermodynamics that describe the effect of pressure, temperature, and salinity on the solubility of oxygen in water, but I have not attempted to deal specifically with the methods by which such effects are measured. Chapter 3 presents the elements of voltammetric analysis which are essential background for a proper understanding of polarographic oxygen sensors, but again there is no description of particular experimental methods. Of course, in presenting basic principles it is not always easy to know at what level the presentation should be pitched, especially in a book such as this which will probably have a very mixed audience, ranging from highly qualified scientists to operating personnel with a limited scientific background. I have chosen to present the thermodynamics and electrochemistry, in general, at a rather elementary level so that those readers who do not have a very great knowledge of these subjects may improve their understanding of them, and therefore of DO measurements. More qualified readers can opt to omit Chapters 2 and 3, although the two chapters might nevertheless serve as a refresher course since, in my experience, the science of thermodynamics is one which is readily forgotten if it is not regularly used, while that of electrochemistry is not regularly used because it is often poorly taught and therefore inadequately learned.

Chapter 4 introduces the Clark-type electrode which I have called a *m*embrane *p*olarographic *o*xygen *d*etector and which, for convenience, I have designated throughout as MPOD, although I realize this is not a common abbreviation. Chapter 3 was necessary to obtain a physicochemical understanding of how a MPOD functions, and Chapter 4 uses this background and extends it to describe this particular system. The mathematics may, at first glance, look formidable, but really only the essential steps are given and very little more than an ability to follow simple algebra is required.

Chapter 5 applies the theoretical treatment of Chapter 4 and looks at the practical aspects of MPOD operation, but again the stress is on principles rather than description. This chapter seeks to explain why the MPOD is such a powerful tool for measuring DO and to discuss the principles of operation so that any user will not only have a better understanding of how the MPOD works, but also will know how to use it more effectively, and will be able to recognize and deal with faults when they occur. Chapter 6 examines a relatively unexplored method of using MPODs—pulsed measurements. These measurements have the advantages of being of a higher sensitivity and of showing less dependence on stirring and temperature than steady-state measurements. And although such measurements are not commonly employed at the moment, in my opinion they will play an increasingly important role in the future.

Perhaps at this point it should be emphasized that although the MPOD does provide a powerful method of analysis, nevertheless it is not a panacea for all the ills of DO measurements. Indeed it does have its shortcomings and problems, as I have tried to indicate in Chapter 5. In addition, as one who has worked for a number of years with electrochemical techniques in general and with electroanalytical methods in particular, I must say that electrochemical systems are often changeful and do not always seem to behave in a logical manner. I am by no means an MCP, as any of my female friends and colleagues would readily testify, but the fact that in both French and German the words for "electrode" and "electrochemistry" are feminine would seem to me to be not without significance. However, to explore this theme any further might lead at best to a misunderstanding and at worst to a lawsuit, so I will let it pass.

Chapters 7 and 8 provide the principles of other electrochemical and nonelectrochemical methods of measurements, respectively. Again there is no umbrella coverage, but rather a critical analysis of the various techniques. When reading a book I personally find it disconcerting to have to consult other texts continually either because I am not familiar enough with the material I am reading and the author has not given enough background for the tyro, or else because I have long since forgotten the fundamentals of what is being discussed. Throughout this book, therefore, I have always tried to present sufficient background information to allow a given principle or technique to be understood, and this is particularly so in these two chapters. Thus, for example, the basis of the electroanalytical use of a dropping mercury electrode can be found in innumerable electrochemical and physical chemical texts; nevertheless I have felt it worthwhile to include a brief summary of the system in Chapter 7 so that the reader, in order to refresh his or her memory, does not have to get up and take down an appropriate volume from the bookshelf, or even have to make a special trip to the library if a suitable text is not at hand. By doing this for the dropping mercury electrode and in other instances I hope the reader's rest in a favorite armchair will thus suffer the minimum of disturbance.

The appendixes serve the traditional function of expanding on points brought up in the main text and of presenting material which would otherwise detract from the main line of argument, while the list of symbols should help the reader to establish quickly the meaning of any given symbol if it is not immediately obvious from the text.

Writing a book is a very personal thing. Here I have tried to give the rationale for what I have written and the way I have written it. The presentation has naturally been influenced by my experience in environmental monitoring, and others with a different outlook would undoubtedly have taken another viewpoint. Nevertheless, as I have mentioned, I have endeavored to make the content of general applicability, and I hope the translation to the jargon and terminology of other disciplines, especially those of medicine and biology, will not prove to be too great a barrier to allow this to be so.

Writing a book may be a very personal thing, but it can never be a completely individual effort. Therefore I should like to thank all those who have helped to make this book possible. The role of Orbisphere Corporation has been indicated elsewhere, but I want to add my own acknowledgment to the company. First to H. Dudley Wright, President of Orbisphere, for it was he who first prompted me to think about writing a text on DO measurements, and who has since that time encouraged and supported me through good periods and bad. Then I thank Dr. John M. Hale, a former colleague, for his stimulating and helpful comments throughout on a wide range of topics, for assistance with the proofreading and indexing, and for providing the results in Appendix H. Last, thanks are due to Frances Stein, whose impeccable typing and layout have helped to bring order out of chaos, and the other secretarial staff of Orbisphere who have had to cope not only with numerous changes and revisions, but also with my cacography without the aid of either a hieroglyphist or a dragoman. I thank Heini Meier for his careful draftsmanship in producing many of the figures from my scruffy sketches, and Drs. Robert T. Jankow and Wolfgang Mehl for many interesting discussions in the early days of my work with DO measurements.

Finally, a special mention for my wife Pauline and my children who must have often wondered why if I was so concerned about the state of the environment I have been stuck in my study writing instead of going out to sample it more often with them.

MICHAEL L. HITCHMAN

Zurich, Switzerland
February 1978

CONTENTS

1

THE NEED AND IMPORTANCE OF DISSOLVED OXYGEN MEASUREMENTS

"Why is this thus? What is the reason of this thusness?"

Artemus Ward

1.1. INTRODUCTION

A good indication of the importance of any subject is the number of papers or articles published on that subject, and not just over a short time interval or from a few authors but consistently over a long period and from numerous authors. Spiehler (1) lists over 600 references directly related to the characteristics, role, and measurement of dissolved oxygen (DO), and this list is only a selection from more than 2000 references in his files. Furthermore, his list of 600 has no reference more recent than 1970, so if DO analysis is typical of most other areas of scientific research, we can probably expect at least 50% more references to be available now. Thus a conservative estimate of the total number of papers directly related to DO measurements would be about 3000, assuming Spiehler's list up to 1970 was complete. However, even if we take this conservative estimate, it is clear that the amount of DO is a rather important parameter for monitoring.

But why is it important to know how much oxygen is dissolved in a liquid? This question can perhaps best be answered by taking a brief look at specific situations where DO is measured.

1.2. NATURAL WATERS

"The water environment can generally be characterised as a dilute aqueous solution containing a large variety of inorganic and organic chemical species, dissolved and in suspension . . ." (2). More specifically, natural waters contain a number of dissolved gases of which oxygen is undoubtedly the most important since it is vital to the health of the flora and fauna in such waters. But the amount of DO in natural waters is not static. It is, rather, in a state of dynamic equilibrium with a delicate balance maintained by biochemical depletion and continued reoxygenation. Thus, bacteria feed on organic matter in an oxidation process which consumes DO, and when

1

there is an ample oxygen and food supply then the bacteria multiply rapidly. This bacterial proliferation is essential, for bacteria are an integral part of the food chain which provides nourishment for fish and other higher forms of life. But the other parts of the chain also consume oxygen, which must therefore be replaced by the twin mechanisms of aeration and photosynthesis if the water is to continue to be living. In rapidly moving streams aeration is generally the most important of the two regenerative processes, but in quiescent lakes and ponds photosynthesis becomes predominant. With aeration of moving waters there is usually little diurnal variation of DO, but when photosynthesis is the main method of replacing oxygen, there is considerable variation since oxygen production is high during daylight hours while at night there is no production.

However, the natural fluctuations in DO can be drastically modified by pollution, even leading to the complete disappearance of the DO in a relatively short time. If waste is discharged into a stream, it lowers the DO level. When the waste load is not too heavy, the DO concentration sags but then recovers. When, however, the waste load is great, then the DO supply may be exhausted, degradation of the aqueous environment begins by anaerobic processes, and the water becomes offensively brackish and septic (3). An example of "oxygen sag" is given in a survey (4) of rivers and streams in Western Pennsylvania which showed that below the outfall of a large sewage treatment plant the concentration of DO fell from 7.1 to 1.2 mg liter^{-1} for a distance of more than half a mile downstream. In this context it is interesting to note that aquatic biologists suggest that for most species of fish the DO concentration should not fall below 5 mg liter^{-1}. Therefore it is important for pollution control engineers to have a continuous record of DO levels in rivers, lakes, and reservoirs.

The oceans are not at present in danger of gross pollution in the way that we have just discussed, but there are many projects being suggested which involve the oceans as dumping grounds for the wastes of modern day society, especially radioactive wastes. Before such plans can be seriously considered, however, the interaction and exchange of deep water and surface water need to be examined and understood. Dissolved oxygen analysis can provide some insight into the problem, and in fact evaluation of the oxygen content data obtained for the North Atlantic Ocean over a period of 25 years suggests that the deep water in this region has not exchanged water with the surface for about 140 years (5). Such information is obviously important to ocean-ographers, waste disposal engineers, and scientific advisers to governments.

1.3. SEWAGE WASTES

In sewage treatment plants biochemical breakdown of the sewage is achieved by bacterial attack in the presence of oxygen. Failure to maintain an adequate supply of air in these processes leads to increased activity of anaerobic, sulfate-reducing bacteria with the resulting production of hydrogen sulfide; this, in addition to being evil smelling, can cause serious and costly corrosion problems (6). On the other hand, aeration is the largest single operating expense, and oxygen levels greater than the optimum concentrations required are wasteful and inefficient. Here too, therefore, DO monitoring is important, and sewage disposal and sanitation engineers need to have a continuous record of the oxygen level at each stage of the treatment process.

1.4. MEDICAL, BIOCHEMICAL, AND MICROBIAL STUDIES

That oxygen is vital for animal and plant life in general is obvious. However, apart from aquatic species, oxygen for life is usually thought of as being needed in gaseous form. But, of course, before a living system can make use of oxygen it must first be dissolved. In the body, whether human or otherwise, oxygen is transported to the cells and tissues by the blood, and clearly a knowledge of DO distribution in blood at various parts of the body can be extremely important for physiological and other medical studies. Various instruments have been described which allow *in situ* monitoring of DO inside the heart and other organs (7, 10), and Spiehler (1), Hahn (8), and Kreuzer and Kimmich (9) list many publications related to oxygen monitoring and respiratory studies.

But oxygen uptake and distribution are only one part of the overall process. Oxygen consumption is equally important, and many studies have been undertaken on the use of DO in cells and tissues (1, 7, 10). In particular, the behavior of tumors and the functioning of various sections of the brain are reflected in the oxygen demand and uptake. For example, a tumor which has outgrown the blood supply has a very low oxygen uptake, but a tumor which has become ulcerated has a large uptake. And DO measurements on interstitial brain fluids can be used to study the effects of various drugs on the brain tissues.

Dissolved oxygen monitoring on tissue in other parts of the body can also provide valuable information. The oxygen tension of the skin of a limb may be used to aid in the determination of the need of amputation in the case of arterial embolism, whereas oxygen measurements on tissue from the liver, kidneys, spleen, various glands, and testes can give insight into malfunctioning, disorders, and effects of drugs (7). Fatt (10) has recently given an extensive review, with a large number of references, of clinical applications of DO measurements, describing in some detail instruments and techniques for monitoring in the circulatory system, in muscle and connective tissue, in malignant tissue, on the skin surface, and in organs.

Photosynthesis, unlike respiration, is not an oxygen consumer but an oxygen generator. Nevertheless, the need to measure DO is still important for, as we mentioned earlier, photosynthesis is one mechanism by which oxygen is replenished in natural waters, and therefore a knowledge of the amount of oxygen produced and its rate of generation can provide valuable insights into the type and quantity of plants needed in the environment. In addition, photosynthesis is a process which is both fascinating in itself and potentially useful outside of plants because of its conversion of carbon dioxide and water, with the aid of sunlight, into sugars and starches. Again, oxygen monitoring could be an aid in understanding the mechanism of this conversion.

Common to both animals and plants are microbial and bacterial reactions, and, as we have noted, these reactions have an important role in the overall cycle of natural processes—and depend on oxygen.

1.5. AGRICULTURE

In emphasizing the important role of plants as photosynthetic agents, supplying and maintaining the oxygen balance in the environment, it must not be forgotten that they need to breathe as well. Plant respiration has therefore also been studied in some detail (1). But respiration in plants is not, unlike photosynthesis, restricted to above ground activity, and most plants show underground respiration as well. Therefore it is essential that a soil is adequately aerated (11). Without sufficient aeration a marked decrease in root activity is found, with mineral salt uptake being affected and the plant ultimately becoming diseased. The chemical composition of soils is also related to oxygen level, with the rate of decomposition of organic materials and the activity of root rotting fungi and other soil fungi being of particular importance.

1.6. FOODS AND DRUGS

The property of oxygen which makes it so essential for most forms of life is its ability to bring about oxidations in living systems, these oxidations being the source of energy for the systems. This oxidizing power can in other circumstances, however, be undesirable and even detrimental. Anyone who enjoys drinking wine knows this; in this case the alcohol in the wine is oxidized to acetic acid or vinegar. Other foodstuffs are also adversely affected by oxidation: milk "goes off"; canned fruit juices show a deterioration in flavor; beer becomes hazy. But these adverse effects do not only occur on storage. In many instances careful oxygen control is necessary during production and preparation (12). Fermentation processes are an obvious example, but perhaps less well known is the need for control in processed food preparation to avoid discolorations, in fruit ripening to obtain an optimum rate of ripening, and in meat curing to produce a desirable flavor, to name but a few (1).

Related to fermentation processes are biosynthetic preparations of drugs and other chemicals. In such syntheses microbial cultures are used to bring about the formation of often highly complex molecules which would be difficult to prepare by more normal chemical techniques. Oxygen is needed in order that the microbial species can respire, but the amount provided must be controlled to prevent unwanted oxidations of the reaction products (12).

1.7. BOILER FEED WATERS

Another oxidation which is highly undesirable is the oxidative corrosion of boiler pipes in steam generators, heaters, and so on. Because of the high temperatures involved even minute traces of DO—of the order of a few parts per billion (ppb) or less—can be extremely detrimental and costly, and rigorous control and monitoring must be maintained to prevent the DO level building up.

1.8. SUMMARY

From this very brief look at specific situations where DO is measured, it is clear there is not a simple answer to our original question about the importance of DO measurements. In some circumstances it is necessary to know the DO level in order to ensure a healthy environment for the flora and fauna. In other cases it is vital as a diagnostic tool in hospitals and surgeries. While in yet other cases it is simply a question of sound economics and good business. And the areas we have used to illustrate the need for DO monitoring by no means exhaust the possibilities. Indeed, there are many, many other areas in the environment, in hospitals, in industry and commerce, in research and development, which use or need some form of oxygen monitoring and which we have not mentioned here. But the fact that there is no simple answer to our original question just emphasizes the importance of DO monitoring—as W. S. Gilbert wrote (13): "Of that there is no manner of doubt, no probable, possible shadow of doubt, no possible doubt whatever."

REFERENCES

1. V. Spiehler, "A bibliography of application areas of Beckman polarographic oxygen analysers," Beckman Applications Research Technical Report No. 545.

2. *Water and Water Pollution Handbook*, Vol I (L. L. Ciaccio, Ed.), Marcel Dekker, New York, 1971.

3. A. V. Kneese and B. T. Bower, *Managing Water Quality: Economics, Technology, Institutions*, John Hopkins Press, Baltimore, Md., 1968.

4. *Manual on Water*, 3rd ed., ASTM Publication, No. 442.

5. L. V. Worthington, "A preliminary note on the time scale in North Atlantic circulation," *Deep Sea Res.*, **1**, 244 (1954).

6. "Hydrogen peroxide control of sewage hydrogen sulphide," FMC Corporation Report 17 J A 53.

7. J. P. Hoare, *The Electrochemistry of Oxygen*, Interscience, New York, 1968.

8. C. E. W. Hahn, A. H. Davis, and W. J. Albery, "Electrochemical improvement of the performance of pO_2 electrodes," *Resp. Physiology*, **25**, 109 (1975).

9. F. Kreuzer and H. P. Kimmich, "Recent developments in oxygen polarography as applied to physiology," in *Measurement of Oxygen* (H. Degn, I. Balslev, and R. Brook, Eds.), Elsevier Scientific Publishing Company, Amsterdam, 1976.

10. I. Fatt, *Polarographic Oxygen Sensors*, CRC Press, Cleveland, 1976.

11. K. A. Smith, R. J. Dowdell, and K. C. Hall, "Measurement of oxygen in the soil atmosphere and in aqueous solutions by gas chromatography," in *Measurement of Oxygen* (H. Degn, I. Balslev, and R. Brook, Eds.), Elsevier Scientific Publishing Company, Amsterdam, 1976.

12. D. E. F. Harrison, "The measurement of dissolved oxygen in continuous fermentations," in *Measurement of Oxygen* (H. Degn, I. Balslev, and R. Brook, Eds.), Elsevier Scientific Publishing Company, Amsterdam, 1976.

13. W. S. Gilbert, *The Gondoliers*.

THERMODYNAMIC ASPECTS
OF DISSOLVED OXYGEN

"A difficulty for every solution."

Attributed to Lord Samuel

The solubility of oxygen in water depends primarily upon three variables: pressure, temperature, and concentration of dissolved salts.

2.1. PRESSURE DEPENDENCE

The solubility of a gas, at constant temperature, can be described by the generalized form of Henry's law (1(a), p. 232):
The fugacity of a solute in dilute solution is proportional to its mole fraction.
The mathematical form of this law is given by the equation:

$$f_2 = kx_2 \tag{2.1}$$

where f_2 is the fugacity of a solute designated by 2, x_2 is the mole fraction of the solute, and k is a proportionality constant.

In order to have a better understanding of what this law is saying we shall digress for a moment to explain the meaning of the various terms—fugacity, mole fraction, and dilute solution.

Fugacity. The concept of temperature as a convenient means of defining thermal equilibrium is a familiar one: thermal equilibrium exists in a system when the temperature is the same in all parts of the system. If there are temperature inequalities in the system then heat will flow in such a direction as to remove these inequalities. This process can be described by saying that heat has a tendency to escape from some parts of the system to other parts, and on this basis temperature is a measure of the escaping tendency. Now the distribution of material substances throughout a system can be considered in a similar way to the distribution of thermal energy. If a substance X is distributed throughout some system then we can speak of the escaping tendency of X in each part, or in each phase, of the system: equilibrium will be achieved when the escaping tendency of each substance in the

7

system is constant throughout the system. As an illustration we can consider the escaping tendency of, say, salt in solution as being more than, equal to, or less than that of solid salt according as the solution is supersaturated, saturated, or unsaturated.

For temperature a number of quantitative scales exist but for convenience only one, in general, is adopted: that based on the perfect gas thermometer. For the analogous concept of escaping tendency applied to the distribution of matter a number of scales also exist, but now it is more convenient to employ different scales of measurement according to the system being studied. Fugacity is just one of these measures of escaping tendency. Vapor pressure can also be regarded as a measure of escaping tendency. The vapor pressure of ice equals the vapor pressure of water at the melting point, but is greater than that of water at all higher temperatures. And in fact the vapor pressure could be quantitatively used for escaping tendency if every vapor behaved as a *perfect gas*; that is, a gas consisting of perfectly elastic molecules, which show no intermolecular interaction, and which, at all temperatures and pressures, obeys the ideal gas equation (1(b), p. 19)

$$PV = RT \tag{2.2}$$

where P is the pressure, V is the molar volume, R is the molar gas constant, and T is the absolute or thermodynamic temperature. Since this condition is not met in reality a different measure of escaping tendency for real, nonideal systems has to be used. Fugacity is this measure. For a vapor which behaves as a perfect gas the fugacity equals the vapor pressure, and so the fugacity may be regarded as a "corrected" vapor pressure. For a real vapor, on the other hand, which does not behave ideally, the fugacity and the vapor pressure are, in general, not the same, although at very low pressures the behavior of real gases and vapors approaches ideality and in this limit the fugacity and vapor pressure do become equal. The dimension of fugacity will obviously be the same as that of vapor pressure, and the metric unit will be the torr, or mm Hg [cf. Appendix A].

Mole Fraction. The composition of a solution can also be expressed by a number of different methods and one very convenient measure is the mole fraction. For a component i, the mole fraction, x_i, of this component is defined as the ratio of the number of moles of component to the total number of moles present. Thus, for a solution containing N_A moles of solvent A, N_B moles of solute B, and N_C moles of solute C, the mole fraction, x_B, of B is given by

$$x_B = \frac{N_B}{N_A + N_B + N_C} \tag{2.3}$$

And, of course,

$$x_A + x_B + x_C = 1$$

From eq. (2.3) we see that in very dilute solution the mole fraction of a solute is proportional to the number of moles of solute in a fixed amount of solvent $(N_A + N_B + N_C \simeq N_A)$, and also is proportional to the concentration, which is the number of moles per unit volume of solution.

Dilute Solution. The reason why "dilute solution" is specified in the statement of Henry's law is that it is only in such a solution that the molecules of solute have just solvent molecules as nearest neighbors. The law applies as long as solute molecules do not approach near enough to each other to interact significantly; only when there is no solute—solute interaction is the escaping tendency strictly proportional to the number of solute molecules in a fixed amount of solvent. In liquids a given molecule will have, in general, about 10 nearest neighbors. So with a mole fraction of 0.1 there is about a 10% chance of a solute molecule finding itself next to another solute molecule, while with a mole fraction of 0.01 there is only a 1% chance; a probability of about 1% is usually small enough to ensure that Henry's law is obeyed.

We now return to the mathematical form of Henry's law

$$f_2 = kx_2 \tag{2.1}$$

This equation shows that for small values of the mole fraction of a dissolved solute (in our case a gas) there is a linear relationship with the fugacity (or the measure of escaping tendency) of the solute. However, the value of the proportionality constant, k, cannot be predicted, for this depends on the nature of both the solvent and the solute, and only in the special case of a solution where there is no solute—solute interaction over the whole range of concentrations (i.e. for $0 \leq x_2 \leq 1$) will the f_2–x_2 curve remain a straight line. In this special case k can be identified with the fugacity of a pure phase of the solute. This is usually given by the symbol f_2° and is referred to as the fugacity of the solute in a standard state.

For a very dilute solution, as we have seen above, the mole fraction of the solute becomes proportional to the concentration. Hence eq. (2.1) becomes

$$f_2 \simeq k'C \tag{2.4}$$

where k' is a new proportionality constant, and C is the concentration of the solute in mol liter^{-1} or g liter^{-1}. But we have also seen that for low pressures fugacity and vapor pressure become identical, and since for many gases, including oxygen, this identity holds very well for partial pressures up to 1 atm or more (3), then for such pressure conditions eq. (2.4) can be approximated to

$$p_2 = k'C \qquad (2.5)$$

where p_2 is the vapor pressure of the gas. That is to say, the vapor pressure of the solute is proportional to the solute concentration in the solution. Or, looking at it from another way, the concentration of gas dissolved in the solution is proportional to the partial pressure* of the gas above the solution.

Equation (2.5) is the original form of Henry's law, which dealt only with gas solubility. We have seen that there exists a more generalized form of the law with a wider applicability, and that this generalized form reduces to the original form under special conditions. If deviations from eq. (2.5) are found then this may be due to the concentration of the dissolved gas being so high that there is a breakdown of the proportionality to the mole fraction, or to the vapor pressure being high enough to differ measurably from the fugacity. It has been reported by Maharajh and Walkley (17) that deviations from Henry's law—even as much as 30%—also arise from mixtures of gases containing oxygen not behaving independently when they are dissolved in water. However, Wilcock and Battino (18) have conclusively shown that while there may be slight deviations from Henry's law they are by no means as great as Maharajh and Walkley claim. Wilcock and Battino suggest the results of Maharajh and Walkley are due to the inherent inaccuracy of the gas chromatographic method of analysis that they used [Section 8.2.2.(e)]. The interaction of oxygen with other gases therefore, under normal conditions, should not be a cause of deviations from eq. (2.5).

For oxygen at a partial pressure of 760 torr the concentration of gas dissolved in water is about 40 mg liter^{-1} at 25°C, which corresponds to a mole fraction of oxygen of about 2×10^{-5}. Thus the solution is very dilute and the approximate eq. (2.4) would be expected to hold. Furthermore, at 760 torr partial pressure there is very little difference between the vapor pressure and fugacity (3), and so eq. (2.5) should hold. Figure 2.1 shows that there is indeed a linear relationship between p_{O_2} and C, and so knowing the slope of the line one can calculate the dissolved oxygen concentration at any partial pressure at the same temperature. For example, for dry air which contains 20.95% of oxygen, and thus which has an oxygen partial pressure of 159 torr (if the total pressure is 760 torr), the concentration of dissolved oxygen obtained either by calculation or by interpolation is 8.6 mg liter^{-1}.

---

* The *partial pressure* of any gas in a mixture of gases is that pressure that the gas would exert if it occupied the entire volume by itself. The total pressure is simply the sum of the partial pressures for an ideal gas mixture (Dalton's law of additive pressures), and even for real gas mixtures this is true provided that the total pressure is not too high.

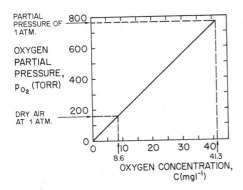

Fig. 2.1 The dependence of oxygen solubility on partial pressure.

2.2. TEMPERATURE DEPENDENCE

All that has been said so far relates to the variation of gas solubility with pressure at constant temperature. But it is well known that, in general, as the temperature of a gaseous solution is raised the gas is driven off until complete degassing occurs at the boiling point of the solvent. The variation of the solubility of a solute with temperature can also be derived from thermodynamic first principles, but we shall be content here, as with Henry's law, merely to state the relationship (1(b), p. 326):

$$\left[\frac{\partial \ln(\bar{x}_2/x_2)}{\partial T}\right]_P = -\frac{\Delta H}{RT^2} \tag{2.6}$$

where x_2 is the mole fraction of the gaseous constituent (denoted by 2) in solution;

$\bar{x}_2$ is the mole fraction of the gas in the vapor above the solution;

T is the thermodynamic temperature;

R is the gas constant (8.314 J K^{-1} mol^{-1});

ΔH is the heat of solution in kJ mol^{-1} at a given temperature and pressure;

P is the total pressure above the solution.

The heat of solution is the heat absorbed or given out when a solute dissolves in a solvent or solution, and as long as the solution remains dilute we need not concern ourselves with a fuller interpretation of this quantity; the matter is discussed in greater depth in standard thermodynamic texts (e.g. 1(a), p. 382). This equation strictly holds only for solutions which behave ideally (i.e. where all molecular interactions are the same), but it is a good approximation to the observed behavior for many dilute solutions.

In the special case where the partial pressure of the gas, and hence its mole fraction in the vapor phase, $\bar{x}_2$, remains constant, eq. (2.6) becomes

$$\left(\frac{\partial \ln x_2}{\partial T}\right)_{P,p_2} = \frac{\Delta H}{RT^2} \tag{2.7}$$

where p_2 denotes a constant partial pressure of the gas. This condition is approximated at a total constant pressure if the vapor pressure of the solvent is negligible in comparison with the gas pressure at all temperatures. Effectively this means that the solvent must be fairly nonvolatile. Failure of this condition or the dilute solution requirement will lead to a breakdown in the relationship.

Over a short temperature range ΔH can also be taken as a constant, in which case simple integration of eq. (2.7) gives

$$\ln \frac{x_2''}{x_2'} = \int_{T'}^{T''} \frac{\Delta H}{R\psi^2} \, d\psi = -\frac{\Delta H}{R}\left(\frac{1}{T''} - \frac{1}{T'}\right)$$

or

$$\ln x_2'' = -\frac{\Delta H}{RT''} + \frac{\Delta H}{RT'} + \ln x_2' \tag{2.8}$$

where the superscript primes indicate two different states of the system. And if the state indicated by a single prime is taken as a reference state, then

$$\ln x_2'' = -\frac{\Delta H}{RT''} + \text{constant} \tag{2.9}$$

Thus a plot of the logarithm of the mole fraction against the inverse of the absolute temperature should be linear, with a slope of $-\Delta H/R$ and an intercept on the temperature axis of $1/T'$, the reference temperature when the mole fraction of solute equals the reference mole fraction, x_2'.

Figure 2.2 shows a plot of eq. (2.9) for oxygen dissolved in water at a total pressure of 760 torr. Two things are immediately apparent from this plot: first that there is a gentle curvature of the line, and second that the gradient is always positive. We could, without making too gross an approximation, draw a straight line through the points, and at any temperature the solubility would be within $\pm 2\%$ of the true value. So to a first approximation eq. (2.9) holds for the solubility of oxygen as a function of temperature in the range $0°C \leqq \theta \leqq 50°C$. However, it is surprising that it holds so well for in the derivation of the equation we assumed that the partial pressure of the gas remains constant; that is, the solvent does not evaporate. Figure 2.2 is a plot based on oxygen solubilities at a fixed total pressure with the air saturated with water vapor. And since the vapor pressure of water in-

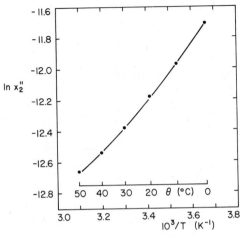

Fig. 2.2 Plot of eq. (2.9).

creases from about 5 torr at 0°C to just over 90 torr at 50°C [see Table 2.4] there is a considerable change in the oxygen partial pressure. Taking a total pressure of 760 torr, then at 0°C the partial pressure of dry air is 755 torr while at 50°C it is 667 torr. So for oxygen, which constitutes 20.95% of dry air, the partial pressure changes from 158 to 139 torr. This means that in a plot of the type in Fig. 2.2 a curved line would be expected with the curvature being in the opposite direction to that actually found, since the partial pressure falls as the temperature rises. The fact that this expected behavior is not found indicates there are other factors controlling the variation of the solubility of oxygen with temperature not considered in this simple approach.

The positive slope means that the heat of solution, ΔH, is negative, or conversely the heat of vaporization is positive. With the general thermo-dynamic rule of negative values for exothermic heat changes and positive values for endothermic changes then a decrease of solubility with increasing temperature is expected, as of course is found.

As the variation of oxygen solubility in water cannot be represented by a simple relationship of the form in eq. (2.9) more complicated expressions which are fitted to empirical data have to be used. The obtaining of this data has occupied a considerable number of workers for over half a century. Both chemical and physical methods have been used in the determinations, and a critical review of the values obtained by these methods, together with suggested "improved" values, is given by Montgomery, Thom, and Cockburn (2). Battino and Clever (3) present all the values obtained by eleven groups of workers up until 1965 in tabular form so that a ready comparison can be made, and we reproduce this data in Table 2.1. The values in this table are not given in the form of solubility in mg liter^{-1} but as the *Bunsen absorption coefficient*, α; this is the volume of gas (reduced to 0°C and 760 torr) which, at the temperature of the measurement, is dissolved in one volume of the solvent when the partial pressure of the gas is 760 torr (1(b), p. 347). The mean value of α at each temperature is also given in the table; the standard error of the mean is never more than $\pm\frac{1}{2}\%$. Whether these mean values, or any particular set of values for that matter, constitute a definitive set of data for oxygen solubility in water is a moot point. However, the

TABLE 2.1 Variation of Oxygen Solubility with Temperature at a Total Pressure
of 1 Atm (3) (Reproduced by Permission of the American Chemical Society)
(Units: Bunsen Coefficient × 10^3)

Temp. °C	0	5	10	15	20	25	30	35	40	50
[1] Winkler	48.89	42.87	38.02	34.15	31.02	28.31	26.08	24.40	23.06	20.90
[2] Fox	49.24	43.21	38.37	34.55	31.44	28.90	26.65	24.85	23.30	20.95
[3] Truesdale, Downing, and Lowden	47.65	41.73	36.98	33.20	30.27	28.00	26.29	24.94	—	—
[4] Elmore and Hayes	49.30	43.15	38.16	34.12	30.88	28.24	25.97	—	—	—
[5] Morrison and Billett	—	—	38.32	34.35	31.13	28.48	26.30	24.48	22.97	20.71
[6] Douglas	—	—	37.97	34.03	30.95	28.30	26.20	—	—	—
[7] Steen	—	42.80	38.37	34.38	31.06	28.65	—	—	—	—
[8] Klots and Benson	—	43.03	38.14	34.23	31.11	28.48	—	—	—	—
[9] Green	49.43	43.31	38.39	34.39	31.12	28.43	26.23	24.40	—	—
[10] Montgomery, Thom, and Cockburn	49.19	43.04	38.13	34.21	31.06	28.49	26.41	24.70	23.31	—
[11] Morris, Stumm, and Galal	49.36	43.33	38.34	34.27	31.03	28.50	26.58	—	—	—
Mean values	49.01	42.94	38.11	34.17	31.01	28.43	26.30	24.63	23.16	20.85

References:

[1] L. W. Winkler, *Ber.*, **22**, 1764 (1889).
[2] C. J. J. Fox, *Trans. Farad. Soc.*, **5**, 68 (1909).
[3] G. A. Truesdale et al., *J. Appl. Chem.*, **5**, 53 (1955).
[4] H. L. Elmore and T. W. Hayes, *J. Sanit. Eng. Div. Am. Soc. Civil Engrs.*, **86**, SA4, 41 (1960).
[5] T. J. Morrison and F. Billett, *J. Chem. Soc.*, 3819 (1952).
[6] E. Douglas, *J. Phys. Chem.*, **68**, 169 (1964); **69**, 1608 (1965).
[7] H. Steen, *Limnol. Oceanog.*, **3**, 423 (1958).
[8] C. E. Klots and B. B. Benson, *J. Mar. Res.*, **21**, 48 (1963).
[9] F. J. Green, Ph.D. Thesis, Massachusetts Institute of Technology, 1965.
[10] H. A. C. Montgomery et al., *J. Appl. Chem.*, **14**, 280 (1964).
[11] J. C. Morris et al., *Proc. Am. Soc. Civil Engrs.*, **85**, 81 (1961).

mean values are probably accurate enough for most, if not all, applications. Green (20) has critically analyzed all results and probable sources of error of all work up to 1965.

The data in Table 2.1 can be described by two forms of equation. The first form is obtained from eq. (2.7) by allowing for the variation of the heat of solution, ΔH, with temperature. If we express this variation in the form of a power series,

$$\Delta H = A' + B'T + C'T^2 + \cdots$$

and as a first approximation neglect all terms beyond the linear one (1(b), p. 77), that is,

$$\Delta H \simeq A' + B'T$$

then substitution in eq. (2.7), together with α for x_2 (1(b), p. 347), and integration gives an equation of the form

$$\ln 10^3\alpha = \frac{A}{T} + B \ln T + C \tag{2.10}$$

where A, B, and C are constants. For oxygen dissolved in water we find by fitting the equation to the mean values of α in Table 2.1 that $A = 8.553 \times 10^3$, $B = 2.378 \times 10$, and $C = -1.608 \times 10^2$ [cf. (19)].

It should be noted that the integration of eq. (2.7) with α instead of x_2 is under the condition of constant partial pressure, which by definition is 760 torr. Also, since α is defined without reference to the total pressure this can be chosen to be constant, so that eq. (2.10) is a thermodynamically accurate way of representing the variation of α with T, at least in so far as our approximate expression for ΔH is correct. We could express $\ln x_2$ in the same form as eq. (2.10) but this would never be strictly thermodynamically correct since, as we have already noted, p_2 does not stay constant.

The other form of equation to describe the variation of α with temperature is obtained by fitting a general power series also to the mean values in Table 2.1:

$$10^3\alpha = a + b\theta + c\theta^2 + d\theta^3 + e\theta^4 + \cdots \tag{2.11}$$

where θ is the temperature in °C. We choose to fit a fourth degree polynomial, and the coefficients calculated by standard curve fitting procedures are given in the first column of Table 2.2; the square of the correlation coefficient is 0.999996. Fox (4) also gave a power series to describe his results and his coefficients are given in the second column of Table 2.2. He did not give a term in θ^4, and calculating α using his coefficients leads to large errors when $\theta > 30$°C. The close correspondence between our coefficients and those of Fox suggests that he in fact fitted a fourth degree polynomial and then omitted to record the last term.

TABLE 2.2　Coefficients of Power Series [eq. (2.11)] for Oxygen Solubility as a Function of Temperature

	Fitted to mean values of Table 2.1	Fox (4)
a	4.900×10	4.924×10
b	-1.335	-1.344
c	2.759×10^{-2}	2.875×10^{-2}
d	-3.235×10^{-4}	-3.024×10^{-4}
e	1.614×10^{-6}	not given

Values of α calculated from eqs. (2.10) and (2.11) for the same temperature agree to within $\pm\frac{1}{2}\%$. Other equations have been given to express the variation of oxygen solubility with temperature (2, 6), but none hold over as wide a range of temperature as the two equations we have given here.

The Bunsen absorption coefficient, while being useful to allow all measurements to be referred to common conditions, is not a very practical measure. Values of α have therefore to be converted to mg liter^{-1}, and the method for doing this is best illustrated by an example.

The value of α, as already mentioned, is the volume of gas, reduced to 0°C and 760 torr, which at the given temperature is dissolved in one volume of solvent. If it is assumed that oxygen behaves as an ideal gas, then, under the reduced conditions of 0°C and 760 torr, one mole of oxygen—that is, 32 g—occupies 22.414 liters. Thus the solubility, S_1, in g cm^{-3}, corresponding to the volume represented by α is

$$S_1 = \frac{32\alpha}{22.414 \times 10^3} \text{ g cm}^{-3}$$

But this is for a *partial* pressure of oxygen of 760 torr, for this is how the Bunsen absorption coefficient is defined. If, however, the total pressure is 760 torr then there will be a contribution to this pressure from the vapor pressure of the water.

From eq. (2.5) in the previous section we see that the solubility, S_2, corrected for this contribution of the water vapor pressure, p_1, to the total pressure is

$$S_2 = S_1 \times \frac{760 - p_1}{760} = \frac{32\alpha}{22.414 \times 10^3} \times \frac{760 - p_1}{760} \text{ g cm}^{-3}$$

Finally, this oxygen solubility is for oxygen saturated water since there is no other gas contributing to the total pressure. Normally the oxygen solubility, S, in air saturated water is required. Dry air consists of 20.95% oxygen and so

$$S = S_2 \times \frac{20.95}{100} = \frac{32\alpha}{22.414 \times 10^3} \times \frac{760 - p_1}{760} \times \frac{20.95}{100} \text{ g cm}^{-3}$$

$$= 3.936 \times 10^{-7} \cdot \alpha \cdot (760 - p_1) \text{ g cm}^{-3} \qquad (2.12)$$

At 25°C, $\alpha = 28.43 \times 10^{-3}$ and $p_1 = 23.76$ torr, and therefore

$$S = 3.936 \times 10^{-7} \times 28.43 \times 10^{-3} \times 736.24$$

$$= 8.24 \times 10^{-6} \text{ g cm}^{-3}$$

$$= 8.24 \text{ mg liter}^{-1}$$

Values calculated from the α values given in Table 2.1 for oxygen solubility in water which is in contact with air saturated with water vapor, as a function of temperature and at a total pressure of 760 torr, are summarized in Table 2.3 and Fig. 2.3.

TABLE 2.3 Solubility of Oxygen in Water in Contact with Water Saturated Air (Total Pressure = 760 torr)

θ (°C)	0	5	10	15	20	25	30	35	40	50
Solubility (mg liter^{-1} or ppm)	14.5	12.7	11.3	10.1	9.1	8.2	7.5	7.0	6.4	5.5

A more complete table with temperature intervals of 1° is given in reference 6. The unit given in Table 2.3 is mg liter^{-1} which, for all practical purposes, is the same as parts per million (ppm) by weight, since water has

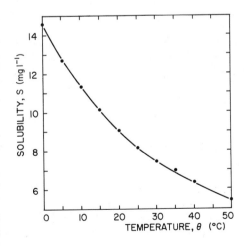

Fig. 2.3 Solubility of oxygen in water in contact with water-saturated air. (Total pressure = 760 torr.)

approximately unit density in the temperature range 0–50°C (5, p. F4). The unit percent saturation is also sometimes used as a measure of oxygen concentration, and any concentration in mg liter^{-1} or ppm can obviously be converted to this unit knowing the value for air saturated water under the same conditions of temperature and pressure.

While discussing ways of expressing measures of DO, it should perhaps be noted that, although we have gone to some length to obtain a measure in terms of concentration, there is in fact no reason why the measure should not be directly given as the partial pressure of gaseous oxygen, p_{O_2}, above the solution, and that is, or could be, in equilibrium with the solution. Indeed, physiologists, biologists, biochemists, and others working in the life sciences usually measure and record dissolved oxygen as a partial pressure in torr or mm Hg, and the measure is often called the *oxygen tension*. This point is discussed further in Appendix C and Section 5.3.3.

2.3. COMBINED DEPENDENCE ON TEMPERATURE AND PRESSURE

To obtain the oxygen solubility at any other barometric pressure use is made of the dependence of the solubility on the oxygen partial pressure, as described previously. If the solubility is S mg liter^{-1} at a barometric pressure of 760 torr (i.e. a partial dry air pressure of $(760 - p_1)$ torr) then the solubility S', at any other total pressure, P_T (i.e. a partial pressure of dry air of $(P_T - p_1)$ torr) is given by

$$S' = S \frac{P_T - p_1}{760 - p_1} \text{ mg liter}^{-1} \tag{2.13}$$

The variation of water vapor pressure with temperature is represented over a short temperature range by (1(b), p. 228)

$$\ln p_1 = -\frac{\Delta H_v}{RT} + \text{constant} \tag{2.14}$$

where ΔH_v is the heat of vaporization, R is the gas constant, and T is the thermodynamic temperature. The constant contains the reference temperature and pressure, and once these have been fixed then, knowing ΔH_v, the vapor pressure at any temperature can be determined. Table 2.4 gives values of water vapor pressure at various temperatures (5, p. D143) and Fig. 2.4 shows the variation for the temperature range 0–50°C plotted according to eq. (2.14). Using this data the solubility at, say, 25°C and a total pressure of 740 torr is given by

$$S' = 8.2 \times \frac{740 - 23.8}{760 - 23.8} = 8.0 \text{ mg liter}^{-1}$$

TABLE 2.4 Variation of Water Vapor Pressure with Temperature

θ (°C)	0	5	10	15	20	25	30	35	40	50
p_1 (torr)	4.6	6.5	9.2	12.8	17.5	23.8	31.8	42.2	55.3	92.5

Combining eqs. (2.12) and (2.13) gives a more general expression for oxygen solubility in terms of the Bunsen absorption coefficient and total pressure:

$$S' = 3.936 \times 10^{-4} \cdot \alpha \cdot (P_T - p_1) \text{ mg liter}^{-1} \qquad (2.15)$$

A grid can obviously be constructed to give the oxygen solubility as a function of temperature and pressure [Appendix B], but a nomogram (7) is quicker and more convenient—Fig. 2.5.

An interesting point, which can be readily seen from the nomogram, is that a change of about 10 torr in the barometric pressure leads to a variation of 0.1 ppm in the dissolved oxygen concentration. This, together with the large temperature dependence [cf. Table 2.3], shows the need of careful measurement of ambient atmospheric conditions. In particular, special notice should be taken of changes in height above sea level when measurements are made. The temperature changes by about 1° for every 100 m change of height, while the variation of atmospheric pressure with altitude is described fairly well by the following equation (10, p. 512):

$$\ln \frac{P_h}{P_r} \simeq \frac{M_A g}{R \delta_T} \ln \frac{T_r - \delta_T h}{T_r} \qquad (2.16)$$

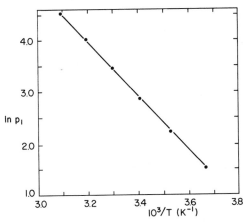

Fig. 2.4 Variation of water vapor pressure with temperature plotted according to eq. (2.14).

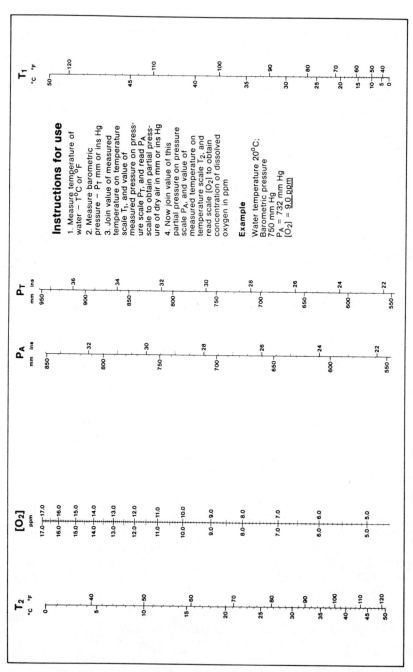

Instructions for use

1. Measure temperature of water – T°C or °F

2. Measure barometric pressure – P_T mm or ins Hg

3. Join value of measured temperature on temperature scale T_1, and value of measured pressure on pressure scale P_T, and read P_A scale to obtain partial pressure of dry air in mm or ins Hg

4. Now join value of this partial pressure on pressure scale P_A, and value of measured temperature on temperature scale T_2, and read scale $[O_2]$ to obtain concentration of dissolved oxygen in ppm

Example

Water temperature 20°C;
Barometric pressure
750 mm Hg
P_A = 732 mm Hg
$[O_2]$ = 9.0 ppm

Fig. 2.5 Nomogram of oxygen solubility in air-saturated freshwater.

where P_h is the atmospheric pressure at some height h above a reference level, P_r is the atmospheric pressure at the reference height, T_r is the thermodynamic temperature at the reference height, M_A is the molecular weight of air (for which we can take a value of 29 g mol^{-1}), g is the gravitational constant, δ_T is the variation in temperature with height, and R is the gas constant.

Since the ratio $\delta_T h/T_r$ is only ~ 0.1 for heights below 4000 m, the log term containing this ratio can be expanded and we can write

$$\ln \frac{P_h}{P_r} \simeq \frac{M_A g h}{R T_r} \qquad (2.17)$$

Then, for example, with a reference pressure of 760 torr at sea level, and with the temperature at this height being 15°C (288 K), the pressure at, say, 600 m above sea level is calculated to be 708 torr. The pressures at various heights, for a sea level temperature of 15°C, are given in Table 2.5.

TABLE 2.5 Variation of
Atmospheric Pressure with
Altitude

Height above sea level (m)	Atmospheric pressure[1] (torr)
0	760
1000	675
2000	599
3000	532
4000	472

[1] Cf. Appendix A.

2.4. DEPENDENCE ON SALT CONCENTRATION

Table 2.6 gives values of the concentration of dissolved oxygen at several temperatures in solutions with various concentrations of Cl$^-$ ion. At any fixed [Cl$^-$] the expected decrease of concentration as the temperature increases is found, but in addition there is a decrease in the oxygen solubility, at a constant temperature, with increased salt concentration. This behavior is characteristic of the solubility of many nonelectrolytes when an electrolyte is added to the solution—it is the phenomenon known as the *salting-out effect*.

TABLE 2.6 Solubility of Oxygen in Water as a
Function of Temperature and Salt Concentration
(Total Pressure = 760 torr)

[Cl$^-$] (g/1000 g)	0	4	8	12	16	20
θ (°C)	Oxygen solubility (mg liter^{-1})					
0	14.5	13.9	13.3	12.6	12.0	11.3
10	11.3	10.8	10.4	9.9	9.5	9.0
20	9.1	8.8	8.5	8.1	7.8	7.4
30	7.5	7.3	7.0	6.7	6.4	6.1

In order to understand this effect better another thermodynamic digression is needed. In the section on pressure dependence of solubility [Section 2.1] we pointed out that fugacity is a measure of the escaping tendency of a real gas and that as the behavior of such a gas approaches ideality so the fugacity and vapor pressure become identical. Since this approach to ideality occurs as the pressure tends to zero we can write

$$\frac{f}{p} \to 1 \quad \text{as} \quad p \to 0$$

The ratio f/p is called the *activity coefficient* and is given the symbol γ_p. This dimensionless ratio expresses the deviation from ideality or the approach to ideal behavior, depending on how one looks at it.

Now it is often convenient or expedient to determine a relative fugacity instead of an absolute value. That is, the ratio between the fugacity of a substance in some given state and the fugacity in some state which, for convenience, is chosen as a standard state. This relative fugacity is called the *activity* and is denoted by "a." Thus the activity is proportional to the fugacity, the value of the proportionality constant depending on the chosen standard state. Representing this constant by $1/k$, then the activity, a_2, of the solute and its fugacity, f_2, may be related by

$$a_2 = \frac{f_2}{k}$$

or

$$f_2 = ka_2 \tag{2.18}$$

In the standard state the activity is obviously unity. It would evidently avoid confusion if one standard state for a given substance was chosen once and for all. So, for example, taking the case of a solvent rather than a solute, liquid water at atmospheric pressure could be chosen to be the standard state. And taking the activity of liquid water as unity, the activity of water in any solution would then be given by the ratio of its fugacity to that of liquid water. However, it is more convenient to be able at any time to choose the standard state best adapted to a particular problem. Considering the case of a solute, such as a gas dissolved in a solvent, the solution is referred to the infinitely dilute solution and it is postulated that the ratio of the activity of the solute to its mole fraction (i.e. a_2/x_2) becomes unity in this infinitely dilute solution at one atmosphere pressure. Thus

$$\frac{a_2}{x_2} \to 1 \quad \text{as} \quad x_2 \to 0$$

and eq. (2.18) can be rewritten as

$$f_2 = kx_2 \quad \text{as} \quad x_2 \to 0 \tag{2.19}$$

This result is identical with Henry's law [eq. (2.1)] and so, on the basis of the standard state used, a_2/x_2 will be unity for all solutions obeying Henry's law. If the solution were ideal over the whole range of concentration then k in eq. (2.19) would be equal to the fugacity of the pure solute at one atmosphere pressure (the standard state), as mentioned previously. The ratio a_2/x_2 can be regarded, just as the ratio f/p is, as a measure of approach to ideality. In other words, it is an activity coefficient, but one based on mole fraction rather than pressure; it is given the symbol γ_x. At infinite dilution γ_x and γ_p are identical and equal to unity, but with increasing concentration they differ.

For a solution composition given in terms of molarity (that is, moles per liter of solution) the standard state is analogous to that proposed above. The activity is defined so that at a given temperature the ratio of the activity of the solute, a_2, to its molarity, C_2, approaches unity in the infinitely dilute solution at one atmosphere pressure. So

$$\frac{a_2}{C_2} \to 1 \quad \text{as} \quad C_2 \to 0$$

The standard state is again a hypothetical one since it corresponds to a one molar solution with properties of the solute corresponding to a solution at infinite dilution. The ratio a_2/C_2 is the activity coefficient based on concentration and is represented by γ_c. As in the previous case, at infinite dilution the activity coefficients, now γ_x and γ_c, are both equal to unity, and for very dilute solutions they are approximately equal. Thus the three coefficients, γ_p, γ_x, and γ_c, all tend to unity at high dilution, but at appreciable concentrations they differ and all deviate from unity.

Now to return to the salting out phenomenon. Consider an aqueous solution which is saturated, and therefore in equilibrium, with a pure, gaseous, nonelectrolyte solute at a given temperature. If an electrolyte is added to this solution then the activity of the gaseous solute is constant provided that the solute is unaffected by the electrolyte. This follows from the definition of activity in eq. (2.18): if the dissolved gas is unaffected by the electrolyte then its escaping tendency from the solution (its fugacity), hence its activity, remains constant.

Let S and S_e represent the solubility of the gas in pure water and in the electrolyte solution, respectively, and let the corresponding activity coefficients be γ and γ_e. Since the standard states remain the same then the activities in the saturated solutions of the gas will be equal, and so

$$S\gamma = S_e\gamma_e$$

or

$$\frac{\gamma_e}{\gamma} = \frac{S}{S_e} \qquad (2.20)$$

From solubility determinations of various nonelectrolyte gases in the presence of electrolytes it has been found (21), and can be verified on a theoretical basis (1(a), p. 584), that $\ln(\gamma_e/\gamma)$ varies approximately in a linear fashion with the ionic strength* of the electrolyte; that is,

$$\ln \frac{\gamma_e}{\gamma} = k_s I \qquad (2.21)$$

The value of the constant k_s, the *salting coefficient* (23), in eq. (2.21) depends on the electrolytic species added and also varies with the nature of the nonelectrolyte solute, but the order for a series of salts is very similar for different nonelectrolytes (23). In aqueous solutions k_s is positive in most cases, that is,

$$\frac{\gamma_e}{\gamma} > 1$$

* The *ionic strength* is given by:

$$I = 0.5 \sum_i C_i z_i^2$$

where C_i is the concentration of ion i, z_i is its charge and the summation is taken over all ions, both positive and negative, in solution. So, for example, a $1M$ solution of NaCl has an ionic strength given by

$$I = 0.5[(1)(1)^2 + (1)(1)^2] = 1$$

A $1M$ solution of $Fe_2(SO_4)_3$, on the other hand, does not have unity ionic strength:

$$I = 0.5[(2)(3)^2 + 3(2)^2] = 15$$

and so from eq. (2.20) it is seen that

$$S > S_e$$

Thus the solubility of a nonelectrolyte is lower in a salt solution than in pure water, as observed, for example, from the figures in Table 2.6 for oxygen.

Substituting for γ_e/γ from eq. (2.20) into eq. (2.21) and considering a uni-univalent electrolyte (e.g. NaCl), where the ionic strength and molarity are identical, we obtain

$$\ln \frac{S}{S_e} = k_s C_\varepsilon \qquad (2.22)$$

where C_ε is the electrolyte concentration. This equation is identical to that originally proposed by Setschenow (21) on empirical evidence. In deriving the equation we have ignored any interaction of the nonelectrolyte gas with itself, which is justified for oxygen at the low solubilities we are considering [Section 2.1]. For gases of high solubility an additional term has to be introduced into eq. (2.22) to take account of such interactions (3).

We can test the equation for the solubility of oxygen. Figure 2.6 shows results at two temperatures and reasonable straight lines are found. The values of oxygen solubility from which Fig. 2.6 is obtained are based on a total pressure of 760 torr and on a chloride concentration in g/1000 g of solution, or ‰. It has been assumed that molarity (g/liter of solution) and molality (g/1000 g of solution) are the same, which is true to within 1% for the conditions we are considering (5, p. D202), and that there is only a very small effect of the salt on the vapor pressure of the water, which is also justified, as Table 2.7 shows. For solutions containing a mixture of electrolytes the concentration, C_ε, in eq. (2.22) should be replaced by the ionic strength,

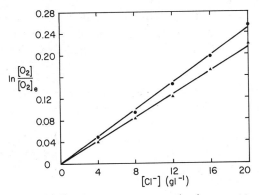

Fig. 2.6 Test of eq. (2.22). $[O_2]$ is the oxygen concentration in pure water, and $[O_2]_e$ that in salt solution. Results at 0°C (●) and at 30°C (▲) are plotted.

**TABLE 2.7   Variation of Solution
Vapor Pressure with Salt Concentration**

[Cl⁻] (g liter⁻¹)	0	9	18	26
θ (°C)	Vapor pressure of solution (torr)			
0	4.6	4.5	4.4	4.4
10	9.2	9.1	8.9	8.8
20	17.5	17.3	17.0	16.7
30	31.8	31.4	30.9	30.4

1. Green and Carritt (22) have shown the validity of the equation to $[Cl^-] >$ 30 g/1000 g.

The mechanism by which the salting out effect works has been the subject of a number of theories (23) and Battino and Clever (3) give a concise summary of these. For our part we can just illustrate the type of mechanism which is occurring by very briefly describing one of the theories. This is the theory based on electrostatic, or coulombic attractive and repulsive, forces and their effect on the dielectric constant of the solvent (i.e. the ratio of the electrical capacity of the solvent to that of air—the lower the constant the better the insulating properties) upon addition of the nonelectrolyte to the salt solution. If the nonelectrolyte decreases the dielectric constant of the aqueous solution, the energy of the nonelectrolyte molecules in the vicinity of an ion will be increased, and so the concentration of the non-electrolyte in the vicinity of the ions is reduced—it is salted out. Conversely, a nonelectrolyte which increases the dielectric constant of the solution is concentrated in the vicinity of the ions and is thus salted in. This theory predicts a linear variation of the logarithm of the activity coefficient of the nonelectrolyte with salt concentration, in agreement with eq. (2.21).

2.5.   DEPENDENCE ON TEMPERATURE, PRESSURE, AND SALT CONCENTRATION

From Fig. 2.6 it is seen that the slope of $\ln[O_2]/[O_2]_e$ versus $[Cl^-]$ is temperature dependent. To take account of this Fox (4) proposed an extension of eq. (2.11) of the form

$$10^3\alpha = a + b\theta + c\theta^2 + d\theta^3 + e\theta^4$$
$$- [Cl^-](p + q\theta + r\theta^2 + s\theta^3 + t\theta^4) \qquad (2.23)$$

The coefficients a–e are those of eq. (2.11) and p–t are new constants. The values of these new constants, obtained by fitting the polynomial to experimental data (2, 4, 24, 25) in the ranges $0 \leq \theta \leq 30°C$ and $0 \leq [Cl^-] \leq 20\%_0$, are given in Table 2.8 together with the constants already determined. To obtain an oxygen solubility from the absorption coefficient the same procedure as described previously is used [Section 2.2].

TABLE 2.8   Values of the Coefficients in eqs. (2.23) and (2.24)

Eq. (2.23)				
	a	4.900×10	p	5.516×10^{-1}
	b	-1.335	q	-1.759×10^{-2}
	c	2.759×10^{-2}	r	2.253×10^{-4}
	d	-3.235×10^{-4}	s	-2.654×10^{-7}
	e	1.614×10^{-6}	t	5.362×10^{-8}
Eq. (2.24)				
	a'	-7.424	p'	-1.288×10^{-1}
	b'	4.417×10^3	q'	5.344×10
	c'	-2.927	r'	-4.442×10^{-2}
	d'	4.238×10^{-2}	s'	7.145×10^{-4}

An alternative equation has been proposed by Green and Carritt (24) which is an integrated form of the van't Hoff equation (1(b), Ch. XIII):

$$10^3\alpha = \exp\left[\left(a' + \frac{b'}{T} + c' \ln T + d'T\right)\right.$$

$$\left. - [Cl^-]\left(p' + \frac{q'}{T} + r' \ln T + s'T\right)\right] \qquad (2.24)$$

The coefficients for this equation are also given in Table 2.8. They are based on measurements for $273.16 \leq T \leq 308.16$ K and $0 \leq [Cl^-] \leq 30\%_0$, and eq. (2.24) is therefore more extensive than eq. (2.23), although in the range of the latter both equations give values of $10^3\alpha$ which agree to better than $\pm 1\%$. The data published by Carpenter (25) for $0 \leq \theta \leq 30°C$ and $0 \leq [Cl^-] \leq 20\%_0$, and which have been extensively tabulated by Gilbert and co-workers (26), also show agreement to within $\pm 1\%$ with values calculated from either of the equations given here.

Instead of chlorinity—the amount of chloride in parts per thousand—which we have used as a measure of the amount of salt in water, the term *salinity* is often used. It is defined by

$$\text{Salinity} = 1.805[Cl^-] + 0.03 \qquad (2.25)$$

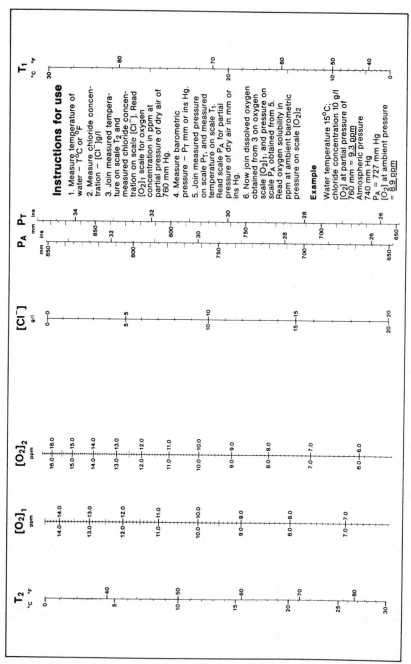

Fig. 2.7 Nomogram of oxygen solubility in air-saturated saline water.

Seawater has a salinity typically of 35 g/1000 g or a chloride content of about 19 g/1000 g (5, p. F145), and therefore falls within the scope of both eqs. (2.23) and (2.24).

Values of oxygen solubility as a function of temperature and chlorinity have been tabulated in a number of places (6, 8, 24, 26) [Appendix B], but again a nomogram (7) is more convenient [Fig. 2.7].

2.6. DEPENDENCE ON DEPTH

Fenn (11) has discussed a series of experiments by Enns et al. (12) in which the effect of high hydrostatic pressure on the partial pressure of oxygen was examined. It was found by extrapolation from the results of these experiments that at a depth of 10 km the partial pressure of oxygen would be about four times that at the surface, but the actual molar concentration would be the same as at the surface. Fenn showed that exponential equations described the variation of both solubility and partial pressure with depth, but he was unable to provide a thermodynamic derivation or a physical explanation of the equations. Subsequently two replies were forthcoming, one by Andrews (13) who developed the equilibrium criteria for solutions in the pressure of a gravitational field, the other by Eckert (14) who based his derivation on the proper definition of the reference state in Henry's law. The two replies are somewhat different but arrive at essentially the same result. We shall only look at Eckert's treatment here since this falls more in line with what we have said so far.

Henry's law is the starting point which is written in terms of natural logs,

$$\ln \frac{f_2}{x_2} = \ln k \qquad (2.26)$$

The Henry's law constant, k, depends, as noted in Section 2.4, on the choice of the standard state. And we saw that for a gas dissolved in a solvent the solution is referred to the infinitely dilute solution. However, a point that was made earlier must now be emphasized; namely, that the standard state is at exactly defined conditions of temperature, T, and pressure, P^0. Moreover, if the pressure P^0 is chosen for convenience, as it often is, to be 1 atm. then one can talk about a standard state. But if, on the other hand, some other value for P^0 is chosen, such as zero pressure or the solvent vapor pressure, then more correctly the term *reference state* should be used.

Now isothermal pressure changes (i.e. pressure changes at constant temperature) can alter the reference condition. This arises because an increase of pressure in a solution results in an increase in the average force acting on the dissolved gas molecules and this tends to push them out of solution. Thus there is an increase in the isothermal work needed to force a solute molecule into solution at elevated pressures, and since the partial pressure, or fugacity, is related to the chemical potential, which is a measure of chemical work [Appendix C], then a correction has to be applied to the reference condition in Henry's law:

$$\ln \frac{f_2}{x_2} = \ln k + \frac{\bar{v}_2^\infty(P - P^0)}{RT} \tag{2.27}$$

where $P - P^0$ is the isothermal pressure change and $\bar{v}_2$ is the partial molal volume* of the gas in the solvent, which at the reference composition is taken as the infinite dilution value, $\bar{v}_2^\infty$. This equation is known as the Krichevsky–Kasarnovsky equation (15), and the second term on the right-hand side is the Poynting correction (13).

The isothermal variation of fugacity in a potential field can be shown to be (16)

$$f_2 = f_2^0 \exp\left(\frac{M_2 g d}{RT}\right) \tag{2.28}$$

where f_2^0 is the fugacity of the solute at the reference pressure P^0, M_2 is the molecular weight of the solute, and g is the acceleration due to gravity. The pressure varies with depth, d, in terms of the density, ρ, of seawater:

$$P = P^0 + \rho g d \tag{2.29}$$

P^0 is the surface pressure which can, for convenience, be taken as the reference pressure, and if it is assumed that at the low surface pressure the ideal gas law is a good approximation for the fugacity of the gas, then combination of eqs. (2.27)–(2.29) gives

$$x_2 = \frac{P^0}{k} \exp\frac{(M_2 - \rho\bar{v}_2^\infty)g d}{RT} \tag{2.30}$$

Equations (2.28) and (2.30) are equivalent to those used by Fenn to calculate the partial pressure and molar concentration as a function of depth. Taking a depth of 10 km then the partial pressure of oxygen at this depth calculated from eq. (2.28) is 3.55 times that at the surface; this is close to the value obtained by Enns by extrapolation of his results. The concentration at 10 km relative to the surface concentration calculated from eq. (2.30) is found to

---

* The *partial molal volume*, $\bar{v}_2$, is the rate of change of the total volume of the solution with the amount of component 2. If we start with a very large amount of solution then $\bar{v}_2$ is the change in its volume when one mole of substance 2 is added; the partial molal volume may be positive or negative.

be 0.97. This is as found by Fenn and arises, as he showed, since M_2 (32 for O_2) is very close to the product of the seawater density ($\rho = 1.023$ g cm^{-3}) and the partial molal volume of oxygen ($\bar{v}_2^{\infty} = 32$ cm^3 mol^{-1}). For other gases greater variations of solubility with depth are predicted.

The physical explanation of the variation of partial pressure and concentration with depth can be obtained by considering eq. (2.30). The argument of the exponential contains two terms which represent two counteracting effects. The term $M_2 gd$ corresponds to the increased potential due to the gravitational force at greater depths; this means that the fugacity or escaping tendency of the gas rises exponentially [eq. (2.28)] because of the increasing weight of a column of gas. The other term accounts for the increased iso-thermal work needed to force a solute molecule into solution at elevated pressures; this gives rise to the Poynting correction as we have already seen.

To summarize: the increased pressure augments the escaping tendency of the gas such that, despite the fact that at 10 km the oxygen solubility is much the same as at the surface, the partial pressure is several times higher than at the surface. Figure 2.8 shows the effect graphically.

From a practical point of view this means that if one is measuring oxygen solubility by a chemical method at great depths then one does not need to worry too much about the effect of hydrostatic pressure. But if a polaro-graphic probe is used, which measures partial pressure rather than concentration [Appendix C], a correction will be needed to convert any reading to solubility; and it should be noted that already at 400 m a correction of 5% is needed—Fig. 2.8.

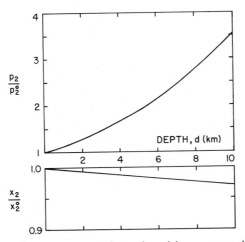

Fig. 2.8 Dependence of oxygen concentration and partial pressure on depth. Note that the ratios of the partial pressure and concentration to their surface values are plotted.

REFERENCES

1. (a) K. S. Pitzer and L. Brewer, *Thermodynamics*, McGraw-Hill, New York, 1961. (b) S. Glasstone, *Thermodynamics for Chemists*, D. Van Nostrand Company, New Jersey, 1947.

2. H. A. C. Montgomery, N. S. Thom, and A. Cockburn, "Determination of dissolved oxygen by the Winkler method and the solubility of oxygen in pure water and sea water," *J. Appl. Chem.*, **14**, 280 (1964).

3. R. Battino and H. L. Clever, "The solubility of gases in liquids," *Chem. Revs.*, **66**, 395 (1966).

4. C. J. J. Fox, "On the coefficients of absorption of nitrogen and oxygen in distilled water and sea water and of atmospheric carbon dioxide in sea water," *Trans. Farad. Soc.*, **5**, 68 (1909).

5. *Handbook of Chemistry and Physics*, 51st ed. The Chemical Rubber Co., Cleveland, 1970.

6. *Standard Methods for the Examination of Water and Wastewater*, 12th ed., Public Health Association Inc., New York, 1965.

7. M. L. Hitchman, "Nomograms for oxygen solubility in fresh and saline waters," to be published.

8. G. C. Whipple and M. C. Whipple, "Solubility of oxygen in sea water," *J. Am. Chem. Soc.*, **33**, 362 (1911).

9. K. H. Mancy, *Instrumental Analysis for Water Pollution Control*, Ann Arbor Science Publishers Inc., Ann Arbor, Mich., 1971.

10. G. R. Noakes, *New Intermediate Physics*, MacMillan, London, 1957.

11. W. O. Fenn, "Partial pressure of gases dissolved at great depth," *Science*, **176**, 1011 (1972).

12. T. Enns, P. F. Scholander, and E. D. Bradstreet, "Effect of hydrostatic pressure on gases dissolved in water," *J. Phys. Chem.*, **69**, 389 (1965).

13. F. C. Andrews, "Gravitational effects on concentrations and partial pressures in solutions: a thermodynamic analysis," *Science*, **178**, 1199 (1972).

14. C. A. Eckert, "The thermodynamics of gases dissolved at great depths," *Science*, **180**, 426 (1973).

15. I. R. Krichevsky and J. S. Kasarnovsky, "Thermodynamical calculations of solubilities of nitrogen and hydrogen in water at high pressures," *J. Am. Chem. Soc.*, **57**, 2168 (1935).

16. K. Denbigh, *The Principles of Chemical Equilibrium*, Cambridge University Press, London, 1966.

17. D. M. Maharajh and J. Walkley, "Lowering of the saturation solubility of oxygen by the presence of another gas," *Nature*, **236**, 165 (1972).

18. R. J. Wilcock and R. Battino, "Solubility of oxygen-nitrogen mixtures in water," *Nature*, **252**, 614 (1974).

19. T. J. Morrison, "The salting-out of non-electrolytes," *J. Chem. Soc.*, **1952**, 3814.

20. E. J. Green, "Redetermination of the solubility of oxygen in sea water and some thermodynamic implications of the solubility relations," Ph.D. Thesis, Massachusetts Institute of Technology, 1965.

21. J. Setschenow, "Concerning the constitution of salt solutions on the basis of their behavior to carbonic acid," *Z. Physik. Chem.*, **4**, 117 (1889).

22. E. J. Green and D. E. Carritt, "Oxygen solubility in sea water: thermodynamic influence of sea salt," *Science*, **157**, 191 (1967).

23. W. L. Masterton, "Salting coefficients for gases in seawater from scaled-particle theory," *J. Soln. Chem.*, **4**, 523 (1975).

24. E. J. Green and D. E. Carritt, "New tables for oxygen saturation of seawater," *J. Mar. Res.*, **25**, 140 (1967).

25. J. H. Carpenter, "New measurements of oxygen solubility in pure and natural water," *Limnol. Oceanogr.*, **11**, 264 (1966).

26. W. Gilbert, W. Pawley, and K. Park, "Carpenter's oxygen solubility tables and nomograph for sea-water as a function of temperature and salinity," Office of Naval Research, Data Report No. 29, 1968.

PRINCIPLES OF VOLTAMMETRY

"And we must take the current when it serves,
Or lose our ventures."

W. Shakespeare, *Julius Caesar*

3.1. INTRODUCTION

In its simplest form an electrochemical cell consists of two electrodes dipping into an electrolyte solution which forms a contact between the electrodes. When such a cell is connected to a current source with a sufficiently high voltage a current will flow through the cell causing an electrochemical reaction to take place at each of the electrodes. This process is called *electrolysis*.

The *cell reaction* is the overall reaction occurring at the electrodes and it is divided into the individual reactions which occur at each electrode. So, for example, a copper and a platinum electrode placed in an acid solution of copper sulfate will show the following reactions on making the copper positive and the platinum negative:

Positive Cu electrode $\qquad Cu \rightarrow Cu^{2+} + 2e^-$

Negative Pt electrode $\quad 2H^+(Pt) + 2e^- \rightarrow H_2$

Overall cell reaction $\qquad Cu + 2H^+ \rightarrow Cu^{2+} + H_2$

In general, in a cell of this type one chooses to study only the reaction occurring at one electrode. This is called the *working* (or indicator) *electrode*. The other electrode is known as the *auxiliary* (or counter) *electrode*, which for the purpose of a given study is important only to complete the circuit. Often a third electrode is introduced, the *reference electrode*. Since it is not possible to measure an absolute value of an electrode potential, potentials have to be measured with respect to a reference level (1, Ch. 1). Typical reference electrodes are the hydrogen electrode, the calomel electrode, and the Ag/AgCl electrode (1).

A current flowing through a cell is characterized by its magnitude and its direction. When positive current flows from the electrode into the electrolyte the electrode acts as an *anode*, and the current is *anodic*. Conversely, a current

34

flowing in the opposite direction is *cathodic* and the electrode is a *cathode*. Obviously in an electrochemical cell one electrode becomes an anode and the other a cathode.

The current, i, passing through a cell in time t seconds causes an electrochemical conversion of w grams of an electroactive species which is proportional to the charge, q ($= it$) coulombs, and to the number of molecular or atomic weights, M, of the substance converted per electron added:

$$w = \frac{Mit}{nF} \tag{3.1}$$

This equation expresses Faraday's first and second laws (2, Ch. 1). n is the number of electrons per electrode reaction and F, the proportionality constant, is called the faraday:

$$1\ F = 96{,}487\ C$$

where C is the symbol for coulombs.

An electrochemical cell connected to a source of electrical power, such as a battery, exchanges energy with the source, thus causing a current to flow. This current is generally a function of voltage applied to the cell and *voltammetry* is the study of these current-voltage characteristics. The term *polarography* is sometimes used interchangeably with the general term *voltammetry*, although more specifically polarography describes the method in which a dropping mercury electrode is used [Section 7.3.1].

In order to discuss voltammetry it is necessary to consider the factors affecting the current which flows. Electrode processes can, in general, be considered as consisting of three consecutive steps: transport of the electroactive species to the electrode, electron transfer, and the removal of reaction products. Of course when electrode dissolution occurs only the last two steps need be considered, and when the electrode process involves a deposition then the first two steps are the relevant ones. But whatever the process, it is clear that the first and third steps involve mass transport and are fundamentally different from the electron transfer, and it is also clear that the overall rate at which an electrode reaction takes place will depend, in the absence of complications, on the relative kinetics of the various steps. Let us look at the processes of electron transfer and mass transport in turn.

3.2.   ELECTRON TRANSFER

Consider a general reaction occurring at an electrode,

$$O + ne^- = R$$

where O is an oxidized species to which an integer number, n, of electrons are added to produce the reduced species R; for the sake of convenience and simplicity we shall assume that both O and R are soluble species. Now, if the rate of the electrode reaction is directly proportional to the concentration of the dissolved species, as it very often is, then one can express this rate in terms of the number of moles of substance, N, transformed at the electrode surface per unit time and per unit area by

$$-\frac{dN_O}{dt} = \frac{dN_R}{dt} = k_f C_0^O - k_b C_0^R \tag{3.2}$$

where the C_0's are the surface concentrations and the k's are the *rate constants* (or, more simply, proportionality constants) for the forward and backward processes. Since any electrochemical reaction takes place at an electrode surface the rate constants used here are *heterogeneous rate constants,* as distinct from homogeneous ones which would be used to describe a reaction occurring in the bulk of the solution. Dimensional analysis [Appendix D] of eq. (3.2) shows that the rate constants have the units of cm s^{-1}. The rate constants can be expressed in terms of the potential at the electrode by (3)

$$k_f = k_f^0 \exp\left[-\frac{\alpha nFE}{RT}\right] \tag{3.3}$$

and

$$k_b = k_b^0 \exp\left[\frac{(1-\alpha)nFE}{RT}\right] \tag{3.4}$$

where E is the electrode potential referred to the normal hydrogen electrode (1, Ch. 2), the k^0's are the values of the rate constants for $E = 0$, F, R, and T have their usual meanings, and α is called the transfer coefficient; the physical significance of α is discussed in Appendix E.

The rate of the electrochemical reaction can clearly be obtained by combining eqs. (3.2)–(3.4), and from this rate the total current is readily obtained by multiplying by the electrode area, A, and by the charge involved in the reduction of one mole of the substance O—eq. (3.1). Thus

$$i = nF\left\{k_f^0 C_0^O \exp\left[-\frac{\alpha nFE}{RT}\right] - k_b^0 C_0^R \exp\left[\frac{(1-\alpha)nFE}{RT}\right]\right\} \tag{3.5}$$

where i is the current density (A cm^{-2}). This equation can be written as:

$$i = i_c + i_a \tag{3.6}$$

where i_c and i_a are the cathodic (corresponding to reduction) and anodic (oxidation) components of the current, and correspond to the first and second terms, respectively, of eq. (3.5).

When the electrode reaction is at equilibrium the cathodic and anodic currents will be equal and opposite and so there will be no net current flow. Setting $i = 0$ in eq. (3.5) and rearranging we obtain

$$\frac{k_f^0 C_\infty^O}{k_b^0 C_\infty^R} = \exp\left(\frac{nFE_e}{RT}\right) \tag{3.7}$$

or

$$E_e = \frac{RT}{nF} \ln \frac{k_f^0}{k_b^0} + \frac{RT}{nF} \ln \frac{C_\infty^O}{C_\infty^R} \tag{3.8}$$

where E_e is the *potential at equilibrium*, and the C_∞'s are bulk concentrations of the electroactive species; since no electrochemical reaction is occurring no net depletion or generation occurs at the surface and thus $C_0 = C_\infty$. The first term on the right-hand side of eq. (3.8) can be represented by E^0 and we then have the classical Nernst equation (1, Ch. 1),

$$E_e = E^0 + \frac{RT}{nF} \ln \frac{C_\infty^O}{C_\infty^R} \tag{3.9}$$

E^0 is the *standard electrode potential* corresponding to equal concentrations of oxidant and reductant. It should be noted that strictly activities [Section 2.4] rather than concentrations should be used in this equation, but our derivation here has used rate constants known as *formal rate constants* which contain activity coefficients within them.

At the standard electrode potential anodic and cathodic currents are equal and so are C_∞^O and C_∞^R. Consideration of eq. (3.5) under these conditions shows that

$$k_f = k_f^0 \exp\left[-\frac{\alpha nFE^0}{RT}\right] = k_b$$

$$= k_b^0 \exp\left[\frac{(1-\alpha)nFE^0}{RT}\right] = k^0 \tag{3.10}$$

where k^0 is a new rate constant. From eqs. (3.10) and (3.5) with a little algebra we obtain

$$i = nFk^0 \left\{ C_0^O \exp\left[-\frac{\alpha nF(E-E^0)}{RT}\right] \right.$$

$$\left. - C_0^R \exp\left[\frac{(1-\alpha)nF(E-E^0)}{RT}\right] \right\} \tag{3.11}$$

This equation shows the current density to be now a function of the difference between the observed electrode potential and the standard electrode potential for the same reaction. However, a more convenient form is where the equation is in terms of $(E - E_e)$ as it is not often that in practice the concentration of oxidant and reductant are equal. This potential difference between the applied, or observed, potential and the equilibrium, or rest, potential is called the *overpotential* and is given the symbol η:

$$\eta = E - E_e$$

Expressing E^0 in terms of E_e from eq. (3.9) and substituting in eq. (3.11) gives

$$i = nFk^0 \left\{ C_0^O \left(\frac{C_\infty^R}{C_\infty^O} \right)^\alpha \exp\left[-\frac{\alpha nF\eta}{RT} \right] \right.$$

$$\left. - C_0^R \left(\frac{C_\infty^O}{C_\infty^R} \right)^{1-\alpha} \exp\left[\frac{(1-\alpha)nF\eta}{RT} \right] \right\} \tag{3.12}$$

Figure 3.1 shows current-overpotential curves corresponding to eq. (3.12) with values of the variables taken as indicated. Also it should be pointed out that it has been assumed that C_0 values are the same as C_∞ values; that is, that there is no change in the reactant concentrations close to the electrode,

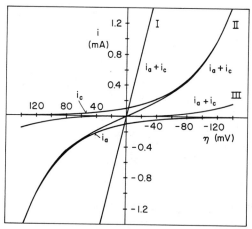

Fig. 3.1 Current-potential curves for charge transfer control. The curves are plotted for the following data:

$$n = 1; \quad T = 298 \text{ K}; \quad A = 1 \text{ cm}^2;$$
$$C_\infty^O = C_\infty^R = 10^{-6} \text{ mol cm}^{-3}; \quad \alpha = 0.5;$$
$$k^0 - \text{Curve I}: 10^{-2} \text{ cm s}^{-1}; \text{Curve II}: 10^{-3} \text{ cm s}^{-1};$$
$$\text{Curve III}: 10^{-4} \text{ cm s}^{-1}.$$

The currents are calculated according to eq. (3.12). The convention chosen of taking cathodic currents as positive and anodic currents as negative is arbitrary but an accepted one. Similarly it is common to plot cathodic currents in the upper right-hand quadrant.

and this will only be justified, as we shall see later, at low overpotentials [Section 3.4].

Considering eq. (3.12) it can be seen that at $\eta = 0$, $i = 0$ and thus there is no net current at the equilibrium potential. On the other hand, at $\eta = 0$ the cathodic and anodic contributions have equal and opposite values:

$$i_a = i_c = i_0 = nFk^0(C_\infty^O)^{1-\alpha}(C_\infty^R)^\alpha = i_0^0(C_\infty^O)^{1-\alpha}(C_\infty^R)^\alpha \tag{3.13}$$

where i_0 is the *exchange current density*, and i_0^0 is the *standard exchange current density*. It should be noted that for eq. (3.13) to be dimensionally correct i_0^0 has the units A $(\text{mol cm}^{-3})^{-1}$ cm^{-2}; in other words, it is the exchange current density corresponding to unit concentration of the reactant species. Although macroscopically a static situation prevails at the electrode surface at $\eta = 0$ (as shown by $i = 0$), on a molecular scale a constant exchange of charge carriers (ions or electrons) takes place through the phase boundary. This exchange corresponds to the equal anodic and cathodic currents, and the current associated with the exchange is called the exchange current: by definition it is positive.

The great importance of the exchange current can be seen if we consider eq. (3.12) for small values of η; that is, $-RT/\alpha nF \ll \eta \ll RT/\alpha nF$. This allows the exponential terms to be expanded and we obtain, on substituting in i_0 from eq. (3.13),

$$i \simeq i_0 \frac{nF\eta}{RT} \tag{3.14}$$

Evidently an ohmic "resistance" of the electrode-solution interface can be defined by

$$R_{ct} = \frac{RT}{nFi_0} \tag{3.15}$$

so that

$$i_0 = \frac{\eta}{R_{ct}} \tag{3.16}$$

In other words, the exchange current is a measure of the ease with which a given electrode reaction occurs. High values of i_0 mean there is a low "resistance" to charge transfer, so-called *reversible systems*, while low values correspond to a high "resistance"—*irreversible systems*. Some typical values of exchange current density are collated in Table 3.1, and from this table it can be seen that i_0 varies not only from system to system but also for the same system on different electrode materials. This has an obvious importance in selecting electrodes for analytical purposes, as will be discussed later [Section 3.5].

TABLE 3.1a   Exchange Currents for Some Redox Reactions (10)

Electrode	Redox system	Indifferent electrolyte	i_0 (A cm^{-2})	α
Pt	$Fe^{3+} + e^- \rightleftharpoons Fe^{2+}$ $[Fe^{3+}] = [Fe^{2+}] = 15$ mM	$1M$ H_2SO_4	5×10^{-3}	0.58
Pt	$Mn^{3+} + e^- \rightleftharpoons Mn^{2+}$ $[Mn^{3+}] = [Mn^{2+}] = 10$ mM	$15N$ H_2SO_4	10^{-5}	0.72
Pt	$Ce^{4+} + e^- \rightleftharpoons Ce^{3+}$ $[Ce^{4+}] = [Ce^{3+}] = 1$ mM	$1N$ H_2SO_4	4.4×10^{-5}	0.75
Hg	$FeOx_3^{3-} + e^- \rightleftharpoons FeOx_3^{4-}$ $[FeOx_3^{3-}] = [FeOx_3^{4-}] = 1$ mM	$0.5M$ K_2Ox	$> 10^{-1}$	—
Hg	$Ti^{4+} + e^- \rightleftharpoons Ti^{3+}$ $[Ti^{4+}] = [Ti^{3+}] = 1$ mM	$1M$ H_2Tar	8.6×10^{-4}	—
Hg	$Cr^{3+} + e^- \rightleftharpoons Cr^{2+}$ $[Cr^{3+}] = [Cr^{2+}] = 1$ mM	$1M$ KCl	9.6×10^{-7}	—

Note: Ox = oxalate; Tar = tartrate.

TABLE 3.1b   Exchange Currents for the Oxygen Electrode Reaction at Various Metals

Electrode	Electrolyte	i_0 (A cm^{-2})	Ref.
Pt	HClO$_4$	10^{-9}	7
Pt black	Acid and alkali	$\sim 5 \times 10^{-10}$	11
Rh	HClO$_4$	6×10^{-9}	7
Ir	HClO$_4$	10^{-11}	7
Au	Alkali	10^{-9}	12

It can also be seen from eq. (3.12) that

$$i = i_0^0 C_0^0 \left(\frac{C_\infty^R}{C_\infty^0} \right)^\alpha \exp\left[-\frac{\alpha nF\eta}{RT} \right]$$

$$\text{for} \quad \eta \gg \frac{RT}{(1-\alpha)nF} \tag{3.17}$$

and that

$$i = -i_0^0 C_0^R \left(\frac{C_\infty^0}{C_\infty^R} \right)^{1-\alpha} \exp\left[\frac{(1-\alpha)nF\eta}{RT} \right]$$

$$\text{for} \quad \eta \gg \frac{RT}{\alpha nF} \tag{3.18}$$

For a simple one electron transfer we take $\alpha = \frac{1}{2}$ (7, Ch. 8), and with $RT/F = 25.6$ mV at 25°C then a value of $\eta = -100$ mV ensures that the total current is almost completely determined by the cathodic contribution, while at $\eta = +100$ mV the anodic branch is dominating; this effect is clearly seen in Fig. 3.1. As the overpotential is increased in either direction (i.e. as one moves further and further away from the rest potential) it is clear that the current will go on increasing exponentially. However, a continually increasing current means that the electroactive species is consumed more and more rapidly at the electrode surface. And eventually a point is reached where the rate of consumption is limited by the rate of transport from the bulk of solution to the electrode. This brings us to the consideration of transport processes.

3.3. MASS TRANSPORT

The transport of a solute in a liquid is governed by three main processes: migration, convection, and diffusion.

3.3.1. MIGRATION

This occurs when a charged particle is placed in an electric field. Thus a negatively charged ion is attracted towards a positive electrode and a positive ion to a negative electrode. For most practical uses of electrochemistry migration is a complicating factor and is suppressed by the addition of an excess of *background*, or *supporting*, *electrolyte*. The electric field in the solution causes a migration of all the ions, those which are not discharged as well as those which are, and so if the concentration of the active ions is low compared with the concentrations of other ions then only a small part of the current in the solution will be due to the migration of the active species.

3.3.2. CONVECTION

This results from movement of the fluid either by forced means (for example, an agitator), or from density or temperature gradients within the fluid—these two forms are generally called *forced* and *natural convection*, respectively. A solution where there is no forced convection is often called a *quiet solution*. Convection causes mass transfer by the moving liquid entraining solute molecules and so transporting them.

Experimental studies of the flow of liquids past surfaces of solid bodies which are wetted by the liquids have shown that the layer of fluid immediately adjacent to the surface remains motionless. Or, in other words, the fluid adheres to the solid surface. Velocity measurements have established that the thickness of this stationary layer is quite small—of the order of several molecular layers—but it is, nevertheless, the absence of slip that arises from this layer which leads to the very important characteristics of fluid flow. On the other hand, at a significant distance from the surface the liquid moves with a velocity comparable to that of the flow of the bulk. Thus, in the zone close to the surface the tangential velocity component undergoes a very abrupt change from a high value at the outer border of the zone to zero at the solid surface. The retardation of the fluid in this *boundary layer* is caused by viscous forces alone, and the velocity gradient that arises in this boundary layer is very large in a direction normal to the surface.

Although the boundary layer occupies only an extremely small volume it exerts a significant influence on the motion of the fluid. The phenomena that take place in the boundary layer are the source of hydrodynamic resistance to the motion of solids through fluids. If the thickness of the *hydrodynamic boundary layer* is designated as δ_0, then for a system with a characteristic length, l, (e.g. the length of a plate, the radius of a tube)

$$\frac{\delta_0}{l} = f(\text{Re}) \tag{3.19}$$

where $f(\text{Re})$ is a function of the Reynolds' number, Re. The Reynolds' number is a dimensionless number which can represent the quantities that are characteristic of a moving liquid [Appendix F].

A detailed analysis (8, Ch. 1) shows that eq. (3.19) can be written as

$$\frac{\delta_0}{l} \sim \frac{1}{\sqrt{\text{Re}}} \tag{3.20}$$

Thus δ_0 is smaller than the characteristic dimension of the body by a factor approximately equal to the square root of the Reynolds' number. For aqueous solutions, typically $\text{Re} \sim 10^3$ and so with $l \sim 1$ cm the thickness of the hydrodynamic boundary layer is $\sim 10^{-2}$ cm.

The concept of a well-defined thickness of the boundary layer does need qualifying. The transition from viscous flow in the boundary layer to inviscid flow in the main stream is smooth and gradual. The quantity δ_0 represents the thickness of the region across which the principal change in velocity from zero to its maximum value occurs. If we arbitrarily define δ_0 as the distance from the surface to the point where the tangential velocity obtains a value equal to 90% of the main stream velocity, v_m, then it can be shown that for flow past a plate (8, Ch. 1)

$$\delta_0 \sim \left(\frac{vx}{v_m}\right)^{1/2} \tag{3.21}$$

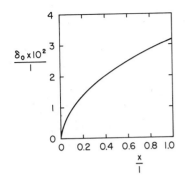

Fig. 3.2 Hydrodynamic boundary layer as a function of the distance along a plate. Values of $v = 10^{-2}$ cm^2 s^{-1}, $v_m = 10$ cm s^{-1} and $l = 1$ cm have been used to calculate the curve. The boundary layer, δ_0, and the distance along the plate, x, have both been divided by the length of the plate, l.

where v is the *kinematic viscosity* [Appendix F] and x is the distance along the plate. Figure 3.2 shows the variation of the boundary layer along a plate.

3.3.3. DIFFUSION

This occurs whenever a concentration gradient (or more strictly, a gradient of chemical potential [Appendix C]) exists, and is the process whereby molecules or ions move from a region of high concentration to one of low concentration: it thus resembles the conduction of heat from a high to a low temperature. From a macroscopic point of view the effect is irreversible unless some external process is used. The quantitative treatment of the phenomenon is based on Fick's law of diffusion which is represented mathematically by

$$J = D \frac{dC}{dx} \qquad (3.22)$$

where J is the one dimensional flow in moles or grams (i.e. the flux) per unit area normal to the direction of flow in unit time, C is the concentration of diffusing species in mol cm^{-3} or g cm^{-3} at a point x cm from some defined origin, dC/dx is the concentration gradient or rate of decrease of concentration per cm in the direction of diffusion, and D is a proportionality constant called the *diffusion coefficient* which by dimensional analysis is seen to have the units of cm^2 s^{-1}. The diffusion coefficient is an exponential function of temperature:

$$D = D_0 \exp\left(-\frac{E_D}{RT}\right) \qquad (3.23)$$

where E_D is the activation energy for diffusion (i.e. the energy required to surmount the barrier that has to be overcome for a species to diffuse from one site to another), D_0 relates to a reference temperature, and R is the gas constant. The important point to note about the temperature dependence of D is that the rate of diffusion increases as the temperature increases; for aqueous solutions D changes by about 2% per degree.

Under normal conditions the value of D of solutes in aqueous solutions is $\sim 10^{-5}$ cm^2 s^{-1} and, as we shall see [Section 3.3.4], this small value, together with the concentration gradients that prevail in unstirred solutions, means that, in general, the rate of diffusion in a static medium is very slow. The school experiment of placing a colored dye in the bottom of a measuring cylinder, carefully adding water and measuring the time taken to obtain a uniform color throughout, illustrates this point. For electroanalytical purposes the higher the flux the higher is the sensitivity of the technique, and since it is only at short times (i.e. just after the current or voltage is applied) that the concentration gradient is high and transport by diffusion is rapid, then in quiet solutions nonsteady state measurements have to be made [Section 3.6]. To obtain a high flux in a steady state situation the solution needs to be stirred so that transport can occur by a combination of convection and diffusion—*convective diffusion*.

3.3.4. CONVECTIVE DIFFUSION

In fluid flow there is a controlling dimensionless parameter, the Reynolds' number. For convective diffusion a similar parameter known as the *Peclet number*, Pe, is used

$$\text{Pe} = \frac{v_m l}{D} \tag{3.24}$$

where v_m is the main stream fluid velocity, l is the characteristic length of the body, and D is the coefficient of diffusion for the diffusing species. This number describes the relationship between convective and diffusional transfer of matter. It plays the same role in convective diffusion as the Reynolds' number plays in fluid flow.

When Pe is small the concentration distribution is determined largely by molecular diffusion: at sufficiently low Pe mass transfer by convection is negligibly small. From the definition of Pe it follows that such a situation occurs, for a given value of D, at sufficiently low liquid velocities and in regions of small dimensions. Conversely, when Pe is large the concentration distribution is determined essentially by convective transfer and molecular diffusion can be neglected. However, in practice for aqueous solutions Pe is $\gg 1$, even for relatively small values of Re [Appendix G], and thus, in general, convective transfer in a liquid is controlling over molecular diffusion In other words, the diffusion coefficient in liquids is so small that, even at low flow velocities, the mass transport by the moving liquid begins to predominate over molecular diffusion.

This situation, however, does not apply near a reaction surface, and this is particularly evident for a system where the condition $C = 0$ must hold at the reaction surface. Near the surface, therefore, a thin liquid layer must exist in which the concentration changes rapidly. The derivative of concentration with respect to distance is in this case very large, with the result that molecular diffusion becomes important despite the small value of D [cf. eq. (3.22)].

Thus at high Pe the liquid may nominally be divided into two regions: (a) a region of constant concentration far from the reaction surface where convection is important, and (b) a region of rapidly changing concentration in the immediate vicinity of the surface where diffusion is important. This division corresponds to the boundary layer treatment as used in hydrodynamics for flows at high Re. Region (b) mentioned here represents a very thin liquid layer which is analogous to the hydrodynamic boundary layer. And just as the effect of viscosity is significant in this hydrodynamic or momentum boundary layer, so molecular diffusion must be taken into account in the thin layer adjacent to the reaction surface, which is therefore called the *diffusion boundary layer*.

The thickness of the hydrodynamic boundary layer, δ_0, is proportional to $v^{1/2}$ [eq. (3.21)]. In diffusion D is analogous to v in convection, although it is numerically one thousand times smaller for aqueous solutions. Thus the thickness of the diffusion boundary layer, δ, should be considerably smaller than that of δ_0. In fact it can be shown that δ and δ_0 are related (8, Ch. 2):

$$\delta \sim \left(\frac{D}{v}\right)^{1/3} \delta_0 \qquad (3.25)$$

For $D \sim 10^{-5}$ cm^2 s^{-1} and $v \sim 10^{-2}$ cm^2 s^{-1}, $\delta \sim 0.1\delta_0$, and so the tangential component of the liquid velocity at the outer edge of the diffusion layer is about 10% of its value far from the surface. Substitution of eq. (3.21) into eq. (3.25) gives

$$\delta \sim D^{1/3} v^{1/6} \left(\frac{x}{v_m}\right)^{1/2} \qquad (3.26)$$

The two layers, diffusion and hydrodynamic, are compared in Fig. 3.3, and either taking $\delta \sim 0.1\delta_0$ or substituting typical values in eq. (3.26) shows that for convective diffusion δ is typically $\sim 10^{-3}$ cm.

Within the diffusion layer the concentration of the solution shows a rapid change and as a first approximation the concentration change may be regarded as linear. Thus the diffusional flux may be approximated by

$$J = \frac{D(C_\infty - C_0)}{\delta} \qquad (3.27)$$

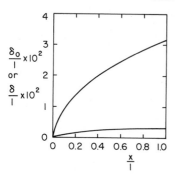

Fig. 3.3 Comparison of hydrodynamic and diffusion boundary layers. $D = 10^{-5}$ cm^2 s^{-1} has been used to calculate the curve for δ/l.

where C_0 and C_∞ are the surface and bulk concentrations, respectively. But it should be noted that, as in the case of the momentum boundary layer, the diffusion layer has no clearly defined boundary. It is simply the region where the concentration gradient is a maximum. Also the fact that δ depends on D for the species which is diffusing means that every species has its own diffusion layer. When several species are diffusing at the same time under given conditions several diffusion layers can exist simultaneously.

3.4. THE CURRENT AT AN ELECTRODE

"The currents measured will therefore be distinctly fluctuating and by no means reproducible."

This quotation is taken from the instructions of an undergraduate experiment at Oxford. The experiment was called "e.m.f's of polarisation" which consisted of placing two copper electrodes in an unstirred solution of copper sulfate and trying to measure the current at an arbitrary time after applying various potentials. The current measured in this experiment can be expressed by the Albery equation (13, Ch. 3),

$$i = f\left(W, N_L, \frac{\partial E}{\partial t}\right) \tag{3.28}$$

where i is the current, W is the number of windows open in the laboratory, N_L is the number of lorries passing at the time, and $\partial E/\partial t$ is a complicated periodic time function describing the movements of the experimenter.

The reasons for the fluctuating currents and the complicated form of eq. (3.28) is that the current at the copper electrode is determined by, among other things, the concentration of the electroactive species (in this case Cu^{2+}) at the electrode surface [cf. eq. (3.5)], and, in general, when a current is passing at an electrode there are local variations in the concentrations of the reactants and products of the electrochemical reaction. The current which is passing destroys some of the electroactive species so perturbing its concentration close to the electrode surface, and in order for a current to continue to flow the electroactive species must be replenished—thus transport of the species from the bulk of the solution to the electrode surface must occur.

As has been seen, where convective diffusion is occurring the solution close to a surface is largely stationary with respect to the surface, and for an electrode process the reactants and products will have to diffuse across this stagnant layer. The main concentration changes occur within this layer and outside it the concentrations approach their value in the bulk of the solution. But in an unstirred solution the thickness of the diffusion layer is not well defined and it can extend out into solution bulk where the local eddies, vibrations, and swirling motions described in eq. (3.28) affect the transport. To prevent these random fluctuations it is necessary that the stagnant diffusion layer close to the electrode be much thinner than the hydrodynamic boundary layer so that chance convective motions in the solution do not become responsible for the transport of species to and from the electrode. This means that with $\delta_0 \sim 10^{-2}$ cm [eq. (3.20) with $l \sim 1$ cm and Re $\sim 10^3$] δ must be $\sim 10^{-3}$ cm—eq. (3.25); electrode systems where this condition is achieved are listed in the next section.

We can now reiterate the main factors which determine the current at an electrode. These are

1. Kinetics of the electrode process.
2. Transport of species to and from the electrode.

There are other factors (6) but they need not concern us here.

The effect of these two factors on the current voltage curve for, say, a redox system can be seen from Fig. 3.4a. This figure shows an experimental i—V curve together with the calculated concentration profiles close to the electrode. An enlarged representation of a profile is given in Fig. 3.4b to show more clearly the way in which the solution is divided into the two regions mentioned above—one where diffusional transport predominates and the other where convective transport is more important. On the current-voltage curve at point C the current is zero, the electrode is at its equilibrium potential—eq. (3.9)—and the concentrations are uniform throughout the solution. At point A, on the other hand, the electrode is very positive and every Fe(II) species that reaches it is rapidly oxidized to Fe(III). As shown the concentration of $Fe(CN)_6^{4-}$ is zero at the electrode surface and the current is determined wholly by the transport of $Fe(CN)_6^{4-}$ from the solution to the electrode. This current is called the *limiting current*. Conversely, at E the electrode is very negative, every Fe(III) is reduced and the limiting

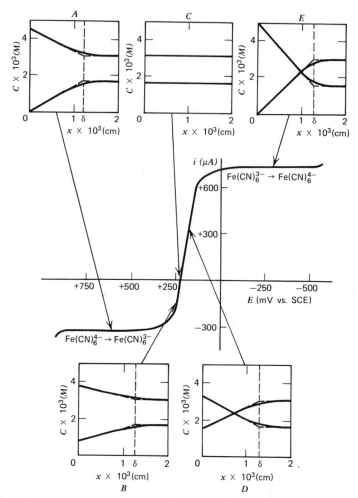

(a) Experimental current—voltage curve and calculated concentration profiles.

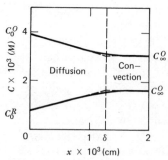

(b) Concentration profile corresponding to point B.

Fig. 3.4 Current—voltage curve and concentration profiles (14). (Reproduced by permission of the Oxford University Press.)

current for the reduction of $Fe(CN)_6^{3-}$ is observed. When in this way the electrode is active enough to remove all the species reaching it the potential of the electrode has no effect on the size of the transport controlled, limiting current. This is the reason why the current does not go on increasing exponentially with potential, as predicted by eq. (3.12), but the characteristic plateaus are obtained. A plateau continues until the electrode potential is sufficient for another electrode reaction to occur, when a similar shaped curve to the first one will be obtained. Points B and D show concentration profiles for cases where the current is controlled partly by the electrode kinetics and partly by transport. These two points are no more than 30–40 mV away from the equilibrium potential and already there is some concentration polarization. Hence the reason for pointing out in connection with Fig. 3.1 that the assumption on which the figure is based, namely that the surface and bulk concentrations are identical, is only valid at very low overpotentials.

3.5. THE IMPORTANCE OF THE LIMITING CURRENT

From eq. (3.22) or (3.27) it is clear that the maximum rate of transport across the diffusion layer is when the concentration gradient is a maximum. Since the bulk concentration of the electroactive species is fixed then this will be when the surface concentration is zero. And, as we have just seen, once the electrode potential reaches a high enough value to reduce the surface concentration to zero then any further increase in the voltage will not lead to an increased flux of material to the electrode.

To convert a flux into terms of a current one simply makes use of Faraday's law to obtain

$$i = nFD \left(\frac{dC}{dx} \right)_{x=0} \tag{3.29}$$

where i is the current density in A cm^{-2}. Thus in order to calculate a current which is determined by diffusion of the electroactive species to or from the electrode first the concentration gradient at the electrode surface must be calculated. This is not always easy, but a general, simplified form of eq. (3.29) can be obtained if the Nernstian approximation of a linear concentration gradient in the diffusion layer is made. Then eq. (3.29) becomes

$$i = nFD \frac{(C_\infty - C_0)}{\delta} \tag{3.30}$$

where C_∞ and C_0 are the concentrations in the bulk of solution and at the electrode surface, respectively. Furthermore, if it is assumed that the thickness of the diffusion layer is given by eq. (3.26) then

$$i \simeq nFD^{2/3}v^{-1/6}v_m^{1/2}x^{-1/2}(C_\infty - C_0) \tag{3.31}$$

And in the special case of the electrode potential being sufficient to maintain the surface concentration at zero this equation reduces to

$$i_L \simeq nFD^{2/3}v^{-1/6}x^{-1/2}v_m^{1/2}C_\infty \qquad (3.32)$$

where i_L signifies a transport limited current.

This equation for the limiting current, while being only approximate, nevertheless shows the right dependence on the variables given, and as such shows the usefulness of limiting current measurements.

First, we see the use in analysis. There is a direct linear relation between the limiting current and the concentration of electroactive species in solution. Second, it can be seen that the sensitivity can be increased in one or more ways:

1. Increase the flow velocity of the fluid.
2. Increase the temperature and so increase the size of the diffusion coefficient and decrease the viscosity of the solution: cf. eqs. (3.23) and (F.3)
3. Reduce the characteristic length of the electrode.

All of these possibilities effectively reduce the thickness of the diffusion layer—see eq. (3.26)—and so increase the rate of transport of material to the electrode, hence augmenting the current. Alternatively, measurements can be made at short times when δ is small and the flux is high—Section 3.6.

One further point should be noted. For a limiting current to be really useful in analysis the plateau should be of the order of several hundred millivolts wide in order that small variations in the control voltage do not lead to different current values. In practice this means that there should be a sharp rise in the current as the applied potential departs from the rest potential and that there should be no additional electrode process occurring which causes the current to rise above the plateau value. This latter condition can be achieved by ensuring that interfering substances are removed from solution before analysis. But even if this is done the decomposition of the solvent can still interfere. For example, with a platinum electrode in neutral aqueous solution water is oxidized at about $+1.0$ V and reduced at ~ -0.8 V with respect to the saturated calomel electrode. This means, of course, that any plateau to be used for analysis must lie in the range $\sim +1.0$ V and ~ -0.8 V. If the end of the plateau is fixed by an interfering electrode process then the plateau should start as soon as possible after departure from the rest potential, which means—see Fig. 3.1—that the charge transfer process should be as rapid as possible. Therefore in selecting an electrode material for a given reaction one must choose the one, all other considerations being equal, at which the reaction has the highest exchange current.

Thus, for example, on this basis platinum would be selected for oxygen reduction rather than iridium [cf. Table 3.1b].

Finally, let us briefly look at some types of electrodes that are suitable for analytical purposes. In order to obtain reproducible currents it is necessary that the transport of the electroactive species from the bulk of the solution to the electrode and vice versa should be well defined and controlled. But this is not all. In order to separate the influence of the different factors not only must there be a defined system of transport, but one should also be able to describe the transport theoretically. Some of the more common types of electrode system are listed in Table 3.2, which also lists the characteristics of each system with respect to time.

TABLE 3.2   Common Types of Electrode System

Electrode	Well-defined transport	Calculable transport	Time characteristics
Stationary	Short times only	Short times only	Transient
Rotating wire	Yes	No	Steady state
Vibrating wire	Yes	No	Cyclic
Mercury drop	Yes	Yes	Cyclic
Rotating disc (R.D.E.)	Yes	Yes	Steady state
Membrane electrode	Yes	Yes	Steady state

The stationary electrode only has defined characteristics for a short time (10^{-1}–10^2 s) after shifting the potential from the zero current point. At longer times the thickness of the diffusion layer becomes too large. No true steady state can be set up [eq. (3.28)], and in all measurements the current, the electrode potential, or both are varying with time. Rotating and vibrating wire electrodes do have well defined diffusion layers, but the thickness of these cannot easily be calculated from first principles, and therefore these electrodes have to be used in a semiempirical fashion. Adams (9) gives a useful discussion of these types of electrode.

The mercury-drop electrode [Section 7.3.1] has been the most widely used system. During the lifetime of each drop it is similar to a stationary electrode with a diffusion layer spreading out from the electrode surface into the bulk of the solution. Before this becomes too large and eq. (3.28) starts to apply, the drop falls off. The current and/or potential vary with time during the life of each drop and the system is therefore a cyclic one. In the rotating disc electrode a known pattern of hydrodynamic flow is imposed on the solution giving well-defined and calculable transport. Several seconds after any change at the electrode a steady state is established and currents and potentials that do not vary with time may be observed. The membrane electrode also has well-defined and calculable transport, and this system will be described in detail in the next chapter.

3.6. TIME-DEPENDENT CHARACTERISTICS

3.6.1. INTRODUCTION

All that has been said so far has referred primarily to steady-state conditions; that is, when stable concentration profiles have been set up and the current has reached a steady value. However, on a couple of occasions it has been indicated that before this situation is reached electrochemical systems show time-dependent characteristics, and, indeed, that these characteristics can be used to advantage in analysis. Now let us look at this situation leading to the steady-state in more detail.

Consider an electrochemical system with no external voltage applied to the cell or no current drawn from it. The potential between the two electrodes will, if one of the electrodes is a reference electrode, obey an equation of the form of eq. (3.9). Such a system can obviously be used as a means of measuring the concentration of an electroactive species, and this method based on the Nernst equation is known as *potentiometry*. The system can now be perturbed from its equilibrium position by an infinite variety of means. For example, a steady dc current could be applied and observations made on how the potential adjusts to this—*chronopotentiometry*. Or one could apply a linearly varying potential with time, cycle between two limits and obtain a *cyclic voltammogram*. A high-frequency sinusoidal voltage or current could be the perturbing force—*oscillographic polarography*. In fact, any conceivable wave form could be used to bring about a potential or current perturbation. And when one couples this with the realization that the perturbation can be such that it takes the electrochemical system into a charge-transfer controlled region, or into a transport controlled region, or into a jointly controlled region, with the additional possibility of either pure diffusional transport or convective diffusion, then it is clear that there is enormous scope for experiment and theory. Fortunately, however, only a few of this multitude of techniques are suitable for routine electroanalysis, and we shall restrict our comments to just two which are particularly pertinent for our needs.

The perturbation we shall be concerned with here occurs when the system is taken from its equilibrium position into the transport controlled region by means of an applied step potential (*potentiostatic* condition) and the

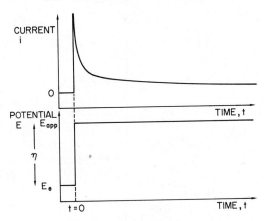

Fig. 3.5 Principle of chronoamperometry.

current variation with time is observed. This technique is known as *chronoamperometry*—Fig. 3.5. In particular we shall deal with the case where this applied potential corresponds to that in the centre of the limiting current region; in other words, where the surface concentration of the electroactive species is zero—$C(x = 0, t > 0) = 0$.

3.6.2. WITH DIFFUSIONAL TRANSPORT ONLY

The current as a function of time is given by eq. (3.29) or eq. (3.30), except that now the concentration gradient, $(dC/dx)_{x=0}$, and the diffusion layer thickness, δ_t, are time dependent. An idea of the form of this time dependence can be obtained by dimensional analysis [Appendix D]. Where there is convection the steady-state diffusion layer thickness can be represented as [cf. eq. (3.26)]

$$\delta = f(D, v, x, v_m) \tag{3.33}$$

Since D is itself a function of v [eq. (G.2)], and since here it is a case where there is no convection, then the time-dependent diffusion layer can probably be written as

$$\delta_t = f(D, t) \tag{3.34}$$

or

$$\delta_t = KD^\varepsilon t^\zeta \tag{3.35}$$

where t is time, K is a constant, and ε and ζ are powers to be determined. Dimensional analysis of eq. (3.35) shows $\varepsilon = \zeta = \frac{1}{2}$ and thus

$$\delta_t = K(Dt)^{1/2} \tag{3.36}$$

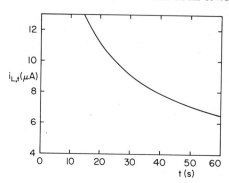

Fig. 3.6 Variation of current with time for $Fe(CN)_6^{4-}$ oxidation in $0.5M$ HCl.

Substitution of this value of δ_t into eq. (3.30) gives

$$i_t = \frac{nF(C_\infty - C_0)}{K}\left(\frac{D}{t}\right)^{1/2} \tag{3.37}$$

And for the condition of applying a potential so that $C_0 = 0$ for all $t > 0$ then

$$i_{L,t} = \frac{nFC_\infty}{K}\left(\frac{D}{t}\right)^{1/2} \tag{3.38}$$

where $i_{L,t}$ represents the time-dependent transport limited current.

Figure 3.6 shows a current time curve for $Fe(CN)_6^{4-}$ oxidation. A useful check on eq. (3.38) is by calculating $i_{L,t}\sqrt{t}$ for various times. This product should be constant for a given system and Table 3.3 gives results for the $Fe(CN)_6^{4-}$ oxidation. In practice, in order to achieve satisfactory results, certain precautions have to be taken to shield the working electrode from thermal convection and other chance motions in the solution (9, Ch. 1).

TABLE 3.3   Constancy of $i_{L,t}t^{1/2}$ for $Fe(CN)_6^{4-}$ Oxidation in $0.5M$ KCl (9)

t (s)	12	18	24	30	36	42	60
$i_{L,t}t^{1/2} \times 10^6$ (As$^{1/2}$)	50.7	50.2	49.9	50.2	50.2	50.4	51.1

Several points should be noted about eq. (3.38) and Fig. 3.6. First, at $t \sim 0$ the current would appear to have an infinite value. This does not occur in practice because, apart from instrumental restrictions, the current is limited at these short times by the electron transfer process. In other words, before the effects of concentration depletion can be "felt" by the electrode, eq. (3.12) momentarily applies. Second, because of the high currents at

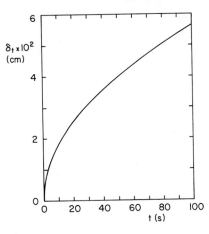

Fig. 3.7 Spread of the diffusion layer for a potentiostatic step in the absence of convection. The curve has been calculated using eq. (3.36) with $D = 10^{-5}$ cm^2 s^{-1} and $K = \sqrt{\pi}$.

short times, measurements made at these times allow, for a given current resolution, smaller concentrations to be determined. Third, at long times $(t \to \infty)$, $i_{L,t} \to 0$ as the diffusion layer spreads out into the solution. In practice this never occurs because, as can be seen from eq. (3.36), for $t > 100$ s, $\delta_t \gg 10^{-2}$ cm, and the transport becomes subject to chance convective motions in the solution [eq. (3.28)]. Figure 3.7 shows the spreading of the diffusion layer out into the bulk of the solution, and Fig. 3.8 the concentration profiles at various times for a substance being removed at the electrode.

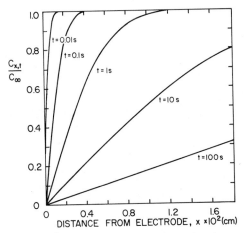

Fig. 3.8 Concentration profiles for a potentiostatic step in the absence of convection. The ratio of the concentration at distance x and time t, $C_{x,t}$, to the bulk concentration, C_∞, is plotted.

3.6.3.   WITH CONVECTIVE DIFFUSION

Under these conditions the equation for the current will have the form

$$i_t = \frac{nFD(C_\infty - C_0)}{\delta} f(t) \tag{3.39}$$

This equation is simply eq. (3.30) multiplied by a function of time which must have the properties

$$f(t) \to 1 \quad \text{as} \quad t \to \infty$$

and

$$f(t) \to \infty \quad \text{as} \quad t \to 0$$

The first condition is necessary since at the steady state the current is that of eq. (3.30); we then have a finite diffusion layer thickness determined by the considerations of Section 3.3.4. The second condition leads to an infinite diffusion current at $t \to 0$, as eq. (3.37) predicts for very short times. At short times the diffusion layer is so thin that the effects of convection are not "felt" by the electrode and essentially the same conditions prevail as in the case of diffusional transport only. This effect can be seen if the currents as a function of time for the two cases—diffusional transport only and convective diffusion—are compared. In Fig. 3.9 $i_{L,t}/i_{L,\infty}$ is plotted as a function of t/τ, where $i_{L,\infty}$ is the value of the current at very long times and τ

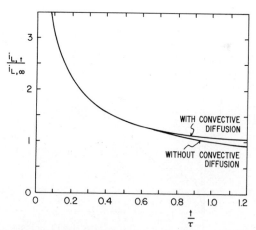

Fig. 3.9 Comparison of the current for a potential step with and without convective diffusion. The curve with convective diffusion is calculated from eq. (3.43) and that without convective diffusion is calculated from eq. (3.42).

is defined as

$$\tau = \frac{\delta^2}{\pi D} \tag{3.40}$$

Equations (3.40) and (3.36) have a similar form if $K = \sqrt{\pi}$ in the latter, which is in fact the value of the constant obtained by a detailed analysis. τ can be thought of as the time for setting up of the steady state diffusion layer in the case of convective diffusion, and with $\delta \sim 10^{-3}$ cm and $D \sim 10^{-5}$ cm^2 s^{-1} then $\tau \sim 100$ ms. In order to plot the current for diffusional transport only, eq. (3.38) is modified to the form

$$i_{L,t} = \frac{nFDC_\infty}{\delta} \left(\frac{\tau}{t} \right)^{1/2} \tag{3.41}$$

so that

$$\frac{i_{L,t}}{i_{L,\infty}} = \left(\frac{\tau}{t} \right)^{1/2} \tag{3.42}$$

The corresponding equation for transport by convective diffusion cannot be obtained by any simple arguments and it is quoted here for the record (5, Ch. 2)

$$\frac{i_{L,t}}{i_{L,\infty}} = 1 + 2 \sum_{m=1}^{\infty} \exp\left(-\frac{m^2 \pi t}{\tau} \right) \tag{3.43}$$

Figure 3.10 shows the concentration profiles at various times for a substance being removed at the electrode.

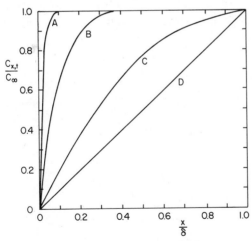

Fig. 3.10 Concentration profiles for a potentiostatic step in the presence of convection. Curves are for the following t/τ values: A: 3.18×10^{-3}; B: 3.18×10^{-2}; C: 3.18×10^{-1}; $D: \to \infty$.

REFERENCES

1. D. J. G. Ives and G. J. Janz, *Reference Electrodes*, Academic, New York, 1961.

2. S. Glasstone, *An Introduction to Electrochemistry*, D. Van Nostrand Company, New Jersey, 1942.

3. S. Glasstone, K. J. Laidler, and H. Eyring, *The Theory of Rate Processes*, McGraw-Hill, New York, 1941.

4. P. Delahay, *New Instrumental Methods in Electrochemistry*, Interscience, New York, 1954.

5. K. J. Vetter, *Electrochemical Kinetics*, Academic, New York, 1967.

6. H. R. Thirsk and J. A. Harrison, *A Guide to the Study of Electrode Kinetics*, Academic, New York, 1972.

7. J. O'M. Bockris and A. K. N. Reddy, *Modern Electrochemistry*, Plenum Press, New York, 1970.

8. V. G. Levich, *Physiochemical Hydrodynamics*, Prentice-Hall, New Jersey, 1962.

9. R. N. Adams, *Electrochemistry at Solid Electrodes*, Marcel Dekker, New York, 1969.

10. J. O'M. Bockris, "Electrode Kinetics," in *Modern Aspects of Electrochemistry*, Vol. 1 (J. O'M. Bockris and B. E. Conway, Eds.), p. 180, Butterworths, London, 1954.

11. W. Vogel and J. Lundquist, "Reduction of oxygen on Teflon bonded platinum electrodes," *J. Electrochem. Soc.*, **117**, 1512 (1970).

12. A. Damjanovic, "Mechanistic analysis of oxygen electrode reactions," in *Modern Aspects of Electrochemistry*, Vol. 5, (J. O'M Bockris and B. E. Conway, Eds.), p. 369, Butterworths, London, 1954.

13. W. J. Albery, *Electrode Kinetics*, Clarendon Press, Oxford, 1975.

14. W. J. Albery and M. L. Hitchman, *Ring-Disc Electrodes*, Clarendon Press, Oxford, 1971.

MEMBRANE-COVERED POLAROGRAPHIC DETECTORS—INTRODUCTION AND THEORY

Moriarty: How are you at Mathematics?
Harry Seacombe: I speak it like a native.

The Goon Show

4.1. INTRODUCTION

It is possible to use any of the electrode systems listed in Table 3.2 for monitoring oxygen, and, indeed, most if not all of them have been used for this purpose at one time or another (1, Ch. VI). However, as pointed out by Clark in his pioneering U.S. patent on a membrane covered electrode (2), all of the systems apart from the membrane electrode suffer from two main drawbacks. First, electroactive or surface active impurities can either react with or poison the sensing electrodes. Second, if measurements are being made *in situ* then changes in the electrolyte composition over a period of time, as for example in a fermentation process, could lead to erroneous results unless frequent calibration checks are made. One way to avoid these effects is to remove the oxygen from the sample by a carrier gas or by vacuum degasification and to measure the oxygen either in the gas phase or by redissolving it in a clean solution [Section 8.2]. But an alternative and, in many ways, more satisfactory approach is to cover the electrodes and electrolyte reservoir with a membrane permeable to oxygen but impermeable to impurities in the test solution. Figure 4.1a illustrates schematically a typical electrode system of this type, and Fig. 4.1b shows several commercial detectors of the same design.

As can be seen from Fig. 4.1a the system consists of a two electrode cell with a membrane separating the electrodes and electrolyte solution from the test solution and, at the same time, keeping a thin layer of electrolyte in direct contact with the cathode. From what has already been said it is clear that the membrane must have a high permeability toward oxygen, and films of PTFE, polyethylene, natural rubber, silicone rubber, and PVC have all been used. The influences affecting the choice of membrane material will be discussed more fully later [Section 5.1.1]. Also we shall deal in detail with the materials that one may use for the electrodes [Section 5.1.2], the composition

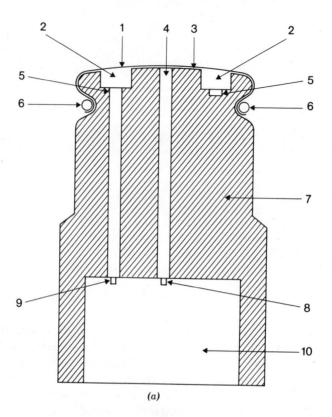

(a)

(1) = Membrane; (2) = Electrolyte reservoir; (3) = Land between cathode and electrolyte reservoir containing thin film of electrolyte solution; (4) = Cathode; (5) = Anode; (6) = 'O'-ring; (7) = Body of detector in insulating material; (8) = Connection to cathode; (9) = Connection to anode; (10) = Space filled with insulating material.

(b)

Fig. 4.1 Membrane-covered polarographic oxygen detectors (MPODs). (*a*) Cross section through a detector. (*b*) Commercial detectors.

of the electrolyte [Section 5.1.3], and how all these factors affect the characteristics of the detector [Sections 5.2 and 5.3]. Here, however, we need merely note the basic principles of operation.

Oxygen from the test solution enters the membrane and diffuses through it into the film of electrolyte solution over the cathode. The dissolved gas diffuses across this layer to the cathode surface where it is reduced:

$$O_2 + 2H_2O + 4e^- \rightarrow 4OH^- \qquad (4.1)$$

The reduction occurs because the cathode is at a sufficiently negative potential with respect to the other electrode, which serves both as an anode and reference electrode. Depending on the electrode materials used either a voltage has to be applied to the cell, as in conventional voltammetric or polarographic analysis (a *voltammetric detector*), or the relative potentials of the two electrodes are such that when an external contact is made between anode and cathode a current flows spontaneously, provided there is oxygen available for reduction (*galvanic detector*); in the latter mode of operation the system is essentially a primary battery or a fuel cell. Whichever system is used though, the cathode is at such a negative potential that it reduces all the oxygen reaching its surface. In other words, the current is diffusion or transport controlled, and it is thus proportional to the oxygen concentration or, more strictly, activity [Sections 2.4 and 3.5 and Appendix C].

The development of membrane covered polarographic oxygen detectors (MPODs) from the early experiments of Davies and Brink (4) with covering the end of a recessed electrode with a membrane of collodion, through the attempt of Kamienski (5) to isolate the surface of his platinum indicator electrode with a semipermeable membrane of silica gel, to the work of Clark (2, 6, 7) with first a cellophane covered electrode and later a polyethylene membrane, has been well reviewed by Hoare (1, Ch. VI). The design of the Clark-type electrode has been the basis of practically all subsequent membrane covered detectors and we reproduce in Fig. 4.2 a drawing from the patent issued to Clark for his device. This drawing apart from its historic interest illustrates once more the principle of a membrane covered detector and also shows a different construction to the cells shown in Fig. 4.1.

Since the Clark patent a number of other devices have been described and some of these will be mentioned in what follows. Reviews of polarographic oxygen detectors have appeared in a number of places (1, 3, 8–10, 14–16) with those of Hoare (1, Ch. VI) and Fatt (16) probably being the most extensive.

Finally, before embarking on a quantitative description the use of the word polarographic should perhaps be clarified. As has been pointed out [Section 3.1] the term *polarography* is usually reserved for electroanalytical techniques based on the dropping mercury electrode, but it is also sometimes

Nov. 17, 1959 L. C. CLARK, JR 2,913,386

ELECTROCHEMICAL DEVICE FOR CHEMICAL ANALYSIS

Filed March 21, 1956

FIG-1

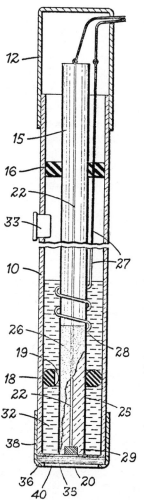

Referring to the drawing, which illustrates a preferred embodiment of the present invention, and particularly to Fig. 1, a typical cell in accordance with the present invention is illustrated as including a tubular body 10 defining a chamber closed at its upper end by a cap 12 and having an insulating rod 15 supported coaxially therewithin by means of an upper washer-like spacer 16, and a lower ring type of spacer 18 slotted at 19 to provide passage for liquid to the lower end of the cell. Within the lower end of rod 15 there is a button-type cathode 20 of a suitable conductive material connected to a cathode lead-in wire 22 which extends within rod 15 to the upper end thereof. The lower outer surface 25 of rod 15 is coated with a suitable conductive material 26 to provide an anode, and an anode lead wire 27 extends through the upper spacer 16 and is wrapped around the lower or coated surface 25 of rod 15 and suitably fixed thereto at a connection 23.

The spacing between the anode and cathode, provided by the thickness of the annular portion 29 of the lower end of rod 15, is thus maintained in predetermined fixed relation, and the space directly below this annular end portion 29 acts as a "bridge" through which ions are transferred and electrical current travels while the reactions occur in the space directly below cathode 20. An electrolyte material 32, preferably including a suitable buffer, is supplied to the well surrounding the lower end of rod 15 and thus to the aforementioned "bridge" and reaction spaces. This well maintains an adequate supply of electrolyte and assures an adequate supply of ions at the appropriate electrode, as necessary for operation of the polarographic cell. If additional solution is required after an extended period of operation, such solution may be added as desired through an opening in the side wall of tube 10, normally closed by a stopper 33.

The electrolyte, the electrodes, and the "bridge" and reaction spaces are all isolated and electrically insulated from the outside of the tube by a selectively permeable barrier means provided by a membrane 35 extending across the end of tube 10 and held in place by an O-ring seal 36 received within the lower cap 38, which in turn is provided with a relatively large central opening 40 to admit the composition to be analyzed to the outer surface of membrane 35. The required spacing between membrane 35 and cathode 20 and the surrounding rod lower end 29 can be provided by roughening the annular lower face of the rod end 29 to provide for access of the electrolyte to the cathode.

The material of which membrane 35 is formed varies in accordance with the properties of the gas, solution, or other composition which it is desired to analyze. For example, when the cell is to be used for determining the oxygen content of a gas or a solution, membrane 35 may be of polyethylene which will pass the oxygen to the interior of the cell, while forming a barrier to other substances which would affect the electrical characteristics of the cell.

Fig. 4.2 The Clark cell (2). (Reproduced with permission.)

used interchangeably with the general term *voltammetry*. Here when we talk of polarographic oxygen detectors is just such a case, and since this description has become embedded in the literature we shall continue to use it.

4.2. THEORY OF THE CURRENT

A mathematical derivation of the equation for the diffusion controlled current at a MPOD has been given by Mancy, Okun, and Reilley (11). The solution for the current as a function of time and its approach to the steady-state value is complicated and we shall only present the problem with its various boundary conditions before giving the equations that result from the theory.

The variation of oxygen concentration with distance from the electrode surface before a potential is applied to a voltammetric detector, or before the external circuit is closed for a galvanic detector, is shown schematically in Fig. 4.3. In the test solution the concentration of oxygen is C_s, and this concentration is constant right up to the outer membrane surface provided the solution is maintained in a homogeneous state by steady stirring. If equilibrium conditions prevail at the membrane—solution phase boundary then at low oxygen concentration a linear relationship exists between the external gas concentration, C_s, and the corresponding equilibrium concentration, C_m, within the membrane:

$$C_m = K_b C_s \tag{4.2}$$

where K_b is the *distribution coefficient*. Equation (4.2) is an expression of the distribution law, which says that a substance distributes itself between two immiscible solvents so that the ratio of its concentration in the two solvents

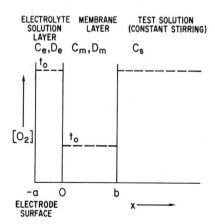

Fig. 4.3 Schematic diagram of initial concentration profile at a MPOD. C_e, C_m and C_s denote the concentration of oxygen in the electrolyte film, the membrane, and the test solution, respectively. D_e and D_m are the diffusion coefficients for oxygen in the electrolyte film and the membrane. t_0 is time zero; that is, before the detector is switched on. (Redrawn from reference 11.)

is approximately a constant, and is equal to the ratio of the solubilities of the substance in each solvent. The membrane is not a material that one normally regards as a solvent, but in this instance it acts as one since the transport of the gas across it involves first the solution of the permeating gas in the membrane material (12, Ch. 2). At the other boundary of the membrane a similar situation to that expressed by eq. (4.2) prevails:

$$C_m = K_0 C_e \tag{4.3}$$

where C_e is the concentration of oxygen in the electrolyte film in front of cathode and K_0 is the distribution coefficient for the electrolyte—membrane interface. For convenience Fig. 4.3 shows C_e and C_s to be equal, in which case obviously $K_b = K_0$, but in practice of course this is not true since some salting out will occur in the electrolyte [Section 2.4].

On closing the cell circuit the potential imposed on the cathode is sufficient to cause immediate and complete reduction of oxygen at the electrode surface—the current is diffusion controlled. As the time of electrolysis increases the concentration profile is continually modified until a steady state with a profile corresponding to $t \to \infty$ is obtained. Figure 4.4 is a schematic diagram showing this variation in the concentration profile. Initially there is a very high flux of oxygen at the electrode surface (i.e. $(\partial C / \partial x)_{x=-a}$ is steep—[Section 3.3.3]), but this flux falls off as the oxygen in the electrolyte layer is consumed. As time progresses depletion of oxygen also occurs in the membrane and this depletion would spread out into the sample if there were no stirring to maintain the bulk concentration, C_s, up to the membrane surface. Since there is stirring though, the oxygen concentration in the membrane at the membrane—sample interface remains constant; that is, $C_m = K_b C_s$ at $x = b$ for all times. At the other boundary of the membrane (electrolyte—membrane) the oxygen concentration, while decreasing as the gas is consumed, nevertheless still obeys eq. (4.3), provided that equilibrium conditions continue to apply to that boundary; in other words, the ratio C_m / C_e remains constant at $x = 0$.

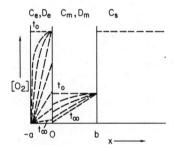

Fig. 4.4 Concentration profiles during attainment of the steady state (11). (Reproduced with permission of Elsevier Sequoia S.A.)

Since the cathode potential is sufficiently negative to reduce the surface concentration of oxygen to zero the current is determined by the rate of diffusion of oxygen to the electrode surface and is given by eq. (3.29), except that the subscript $x = 0$ is changed to $x = -a$, and D and C are subscripted with e in order to conform to the nomenclature used in Fig. 4.4:

$$i = nFD_e \left(\frac{\partial C_e}{\partial x} \right)_{x = -a} \tag{4.4}$$

It is thus necessary to calculate the concentration gradient at the electrode, and this Mancy and co-workers did with the following assumptions:

1. The oxygen only enters the cell through the membrane in a direction perpendicular to the electrode surface and a negligible amount enters from the edges.
2. The diffusion coefficients, D_e and D_m, for the oxygen in the electrolyte film and membrane, respectively, are constants independent of concentration, time, and site in the two phases.
3. Concentrations are used instead of chemical potentials or activities.

The first assumption is a reasonable one provided the electrolyte film and membrane are not too thick. If the coefficient of diffusion is the same for diffusion normal to and diffusion by paths nonnormal to the plane of the electrode surface, then nonnormal diffusion only becomes important when the electrolyte film and membrane are thick—cf. eq. (3.36). Since the thickness of film and membrane are typically ~ 20 μm then access of oxygen to the electrode by nonperpendicular diffusion will be limited to a region ~ 0.02 mm around the edges. To a first approximation this means that the assumption will be valid for an electrode with radius greater than ~ 0.2 mm, as the amount of oxygen having access by normal diffusion relative to that having access by nonnormal diffusion will be in proportion to the ratio of the electrode area to the surrounding annulus—$\pi r^2 / 2\pi \, \Delta r \simeq r / \Delta r \gg 1$ for $r = 0.02$ cm and $\Delta r = 0.002$ cm.

The second assumption is a common one, and is a good one provided there are no gross inhomogenities in the phases or significant changes in composition with time (13, Ch. 1). The third assumption is made in order to simplify the presentation, and its implications will be discussed later [Section 5.3.3].

With these assumptions, together with the conditions

$$
\begin{aligned}
C_e &= 0 & \text{at } x = -a, \quad & t > 0 \\
C_m &= K_0 C_e & \text{at } x = 0 & \\
C_m &= K_b C_s & \text{at } x = b, \quad & t \geqq 0
\end{aligned}
$$

(the last two of these three conditions being based on an assumption of equilibrium conditions prevailing at the membrane—solution interfaces), the concentration gradient at the electrode surface is calculated. However, because of the complex nature of the general solution a number of particular solutions are obtained for $(\partial C_e/\partial x)_{x=-a}$, depending on special conditions. These special conditions correspond to the relative times of diffusion of the oxygen across the membrane and electrolyte layer and to the time that the detector has been switched on. To compare the times of diffusion eq. (3.36) is used:

$$\delta_t = K(Dt)^{1/2} \tag{3.36}$$

Putting this into the nomenclature we are using now shows that $a/\sqrt{D_e}$ has to be compared with $b/\sqrt{D_m}$. Mancy and co-workers give solutions for $(\partial C_e/\partial x)_{x=-a}$ for the three cases

$$\text{(a)} \quad \frac{a}{\sqrt{D_e}} \gg \frac{b}{\sqrt{D_m}}$$

$$\text{(b)} \quad \frac{a}{\sqrt{D_e}} = \frac{b}{\sqrt{D_m}}$$

$$\text{(c)} \quad \frac{a}{\sqrt{D_e}} \ll \frac{b}{\sqrt{D_m}}$$

The first case corresponds to the situation when diffusion in the electrolyte film is rate determining, the third case to diffusion in the membrane being governing, and the middle case when both diffusion processes are controlling. When the detector is first switched on the diffusion layer will be very thin and confined to the electrolyte layer [cf. Fig. 4.4 with Figs. 3.7 and 3.8]. This obviously means that the membrane cannot have an effect on the overall transport process, and so at very short times the solution for the concentration gradient obtained with case (a) must be used. As the diffusion layer spreads out it eventually reaches the interface between the membrane and electrolyte film; at this point there is joint transport control. Finally, the diffusion layer passes into the membrane, and provided that $a \ll 10b$ and $D_e \gg D_m$ then the solution corresponding to case (c) need only be considered. In practice, $a \simeq b$ and $D_m \sim 10^{-2}D_e$ [Section 5.1.1] so that the inequality of condition (c) is only just beginning to hold. For simplicity, however, we shall assume here that it does hold and obtain the expression for the steady-state current on this assumption. Appendix H deals with the influence of the electrolyte layer thickness on the steady-state current. The case of joint control under transient conditions is very complicated and we shall not concern ourselves with it. The current at very short times is of interest though [Chapter 6] and so we shall also consider the solution for case (a).

Thus there are two time domains to deal with—very short times and longer times. However, three solutions for $(\partial C_e/\partial x)_{x=-a}$ still need to be considered: one solution which holds at very short times when diffusion in the electrolyte film is controlling, and two solutions for when diffusion in the membrane is controlling. It is convenient, because of the mathematical complexity, to have two solutions for the latter case, one for short time intervals and the other for longer times as the detector approaches its steady-state condition.

At very short time intervals before the diffusion layer reaches the membrane

$$\left(\frac{\partial C_e}{\partial x}\right)_{x=-a} = \frac{C_{e,t=0}}{(\pi D_e t)^{1/2}}\left[1 + 2\sum_{n=1}^{\infty}\exp\left(-\frac{n^2 a^2}{D_e t}\right)\right] \tag{4.5}$$

At short times when transport in the membrane is controlling

$$\left(\frac{\partial C_e}{\partial x}\right)_{x=-a} = \frac{C_{e,t=0}K_0}{D_e}\left(\frac{D_m}{\pi t}\right)^{1/2}\left[1 + 2\sum_{n=1}^{\infty}\exp\left(-\frac{n^2 b^2}{D_m t}\right)\right] \tag{4.6}$$

At longer times after switching on the detector

$$\left(\frac{\partial C_e}{\partial x}\right)_{x=-a} = \frac{C_{e,t=0}D_m K_0}{bD_e}\left[1 + 2\sum_{n=1}^{\infty}\exp\left(-\frac{n^2\pi^2 D_m t}{b^2}\right)\right] \tag{4.7}$$

The concentration of oxygen in the electrolyte film before the detector is switched on is clearly related to the oxygen concentration in the test solution by eqs. (4.2) and (4.3):

$$C_{e,t=0} = \frac{K_b}{K_0} C_s \tag{4.8}$$

The diffusion of gas molecules in the membrane can be expressed in terms of Fick's law [eq. (3.22)]:

$$J_m = D_m \frac{C_{m,x=b} - C_{m,x=0}}{b} \tag{4.9}$$

where $C_{m,x=b}$ and $C_{m,x=0}$ are respectively the oxygen concentrations within the membrane at the test solution and electrolyte interfaces. This flux, J_m, can also be described in terms of the *permeability* of the membrane to the gas (12):

$$J_m = P_m \frac{C_s - C_{e,x=0}}{b} \tag{4.10}$$

where P_m is the permeability coefficient of the membrane. From eqs. (4.2), (4.3), (4.9), and (4.10) a relationship is obtained between the permeability coefficient and diffusion coefficient:

$$P_m = D_m \frac{K_b C_s - K_0 C_{e,x=0}}{C_s - C_{e,x=0}}$$

which at longer times when $C_{e,x=0} \ll C_s$ becomes

$$P_m = K_b D_m \qquad (4.11)$$

Substituting from eqs. (4.8) and (4.11) into eqs. (4.5)–(4.7) and then, in turn, into eq. (4.4) the expressions for the detector current are obtained. At very short times

$$i_{L,t} = nF \frac{K_b}{K_0} C_s \left(\frac{D_e}{\pi t}\right)^{1/2} \left[1 + 2 \sum_{n=1}^{\infty} \exp\left(-\frac{n^2 a^2}{D_e t}\right)\right] \qquad (4.12)$$

At short times when the diffusion layer has entered the membrane

$$i_{L,t} = nF C_s \left(\frac{P_m}{\pi t}\right)^{1/2} \left[1 + 2 \sum_{n=1}^{\infty} \exp\left(-\frac{n^2 b^2}{D_m t}\right)\right] \qquad (4.13)$$

At longer time intervals

$$i_{L,t} = nF C_s \left(\frac{P_m}{b}\right) \left[1 + 2 \sum_{n=1}^{\infty} \exp\left(-\frac{n^2 \pi^2 D_m t}{b^2}\right)\right] \qquad (4.14)$$

The thickness of the electrolyte film, a, is generally of the order of 10 μm [Section 5.1.1] and, as has been seen previously [Section 3.3.3], diffusion coefficients in aqueous solutions are typically $\sim 10^{-5}$ cm^2 s^{-1}. Therefore on the basis of eq. (3.36) it is clear that the diffusion layer will have spread to the electrolyte—membrane interface within the region of a few hundred milliseconds. And so eq. (4.12), since it is only applicable when the diffusion layer is confined to the electrolyte film, will hold for $t < \sim 0.1$ s. Substituting this maximum value of t, together with the typical values of a and D_e and the minimum value of n, shows that at times < 0.1 s the summation in eq. (4.12) can be neglected and the equation reduces to

$$i_{L,t} = nF \frac{K_b}{K_0} C_s \left(\frac{D_e}{\pi t}\right)^{1/2} \qquad \text{for } t < \sim 0.1 \text{ s} \qquad (4.15)$$

This is of the same form as eq. (3.38) for the time-dependent, limiting current when diffusional transport only occurs, which is as one would expect since the transport problem is identical to that discussed previously [Section 3.6.2]; K_b/K_0 expresses the effect of the salting out in the electrolyte.

Equation (4.13) can also be reduced to a simple form. Typical values of b and D_m are 20 μm and 10^{-7} cm^2 s^{-1}, respectively [Section 5.1.1], and for $t \lesssim \sim 10$ s the exponential term is small for all n and so

$$i_{L,t} = nF C_s \left(\frac{P_m}{\pi t}\right)^{1/2} \qquad \text{for } t < \sim 10 \text{ s} \qquad (4.16)$$

After $t \sim 10$ s the diffusion layer will be approaching the outer face of the membrane [eq. (3.36)] and the current will be approaching its steady-state value. From hereon, therefore, eq. (4.14) applies. The similarity between eqs. (4.15) and (4.16) will be noted. This similarity arises because in both cases it is diffusional transport alone which is controlling. The changeover from the diffusion coefficient in the electrolyte to the permeability coefficient in the membrane reflects the change of the medium governing the transport.

Equation (4.14) then applies for $t > \sim 10$ s. The term in brackets will be recognized as being of the same form as the term that was introduced in the expression for the current in order to take account of transport by convective diffusion—eq. (3.43). In convective diffusion, it will be recalled, the diffusion layer is restricted to a finite thickness [Section 3.3.4] and this is essentially what is happening here since at the outer edge of the membrane diffusional transport is not allowed to spread into the bulk of the solution because of the well-stirred test solution. The membrane thus acts as a diffusion layer of finite thickness and therefore one would expect a similar equation to that obtained before, the only differences being that parameters characteristic of a membrane (D_m and P_m) rather than of an electrolyte solution are used.

The steady-state current is obtained from eq. (4.14). The exponential term becomes negligible when the argument is ≥ 5. With $b \sim 20$ μm and $D_m \sim 10^{-7}$ cm^2 s^{-1} this corresponds to $t \sim 20$ s. Thus

$$i_L = nFC_s \frac{P_m}{b} \qquad \text{for } t > \sim 20 \text{ s} \qquad (4.17)$$

As mentioned earlier, this derivation of the steady-state current assumes only transport in the membrane to be of importance, and for most practical detector designs this is approximately true. If the electrolyte layer becomes too thick though the approximation is no longer valid; Appendix H examines the case when the approximation breaks down. Figure 4.5 shows the variation

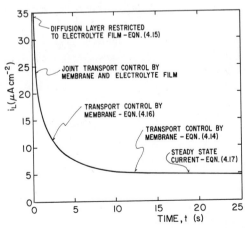

Fig. 4.5 Variation of current with time for a MPOD.

of current with time and the sections corresponding to the equations discussed here are identified.

REFERENCES

1. J. P. Hoare, *The Electrochemistry of Oxygen*, Interscience, New York, 1968.
2. L. C. Clark, "Electrochemical device for chemical analysis," U.S. Patent 2,913,386, issued Nov. 17, 1959.
3. K. H. Mancy, *Instrumental Analysis for Water Pollution Control*, Ann Arbor Science Publishers Inc., Ann Arbor, Mich., 1971.
4. P. W. Davies and F. Brink, "Microelectrodes for measuring local oxygen tension in animal tissues," *Rev. Sci. Instr.*, **13**, 524 (1942).
5. B. Kamienski, "An electrical microcell for the detection of small concentration changes in acid or base of oxidising and reducing substances in non-conducting liquids and gases," *Bull. Intern. Acad. Polon. Soc.*, **1949A**, 81.
6. L. C. Clark et al., "Continuous recording of blood oxygen tensions by polarography," *J. Appl. Physiol.*, **6**, 189 (1953).
7. L. C. Clark, "Monitor and control of blood and tissue oxygen tensions," *Trans. Am. Soc. Artificial Internal Organs*, **2**, 41 (1956).
8. H. Noesel, "Modern technology for electrochemically measuring dissolved oxygen in aqueous media," *Messtechnik*, **81**, 15 (1973).
9. J. Izydorczyk and W. Misniakiewicz, "Electrochemical detectors for continuous determination of oxygen," *Chemik.*, **26**, 114 (1973).
10. V. Spiehler, "A bibliography of application areas for Beckman polarographic oxygen analysers," Beckman Technical Report No. 545.
11. K. H. Mancy, D. A. Okun, and C. N. Reilley, "A galvanic cell oxygen analyser," *J. Electroanal. Chem.*, **4**, 65 (1962).
12. J. Crank and G. S. Park, *Diffusion in Polymers*, Academic, New York, 1968.
13. J. Crank, *The Mathematics of Diffusion*, Oxford University Press, Oxford, 1956.
14. "Instrumentation for Environmental Monitoring - Water," compiled by Lawrence Berkeley Laboratory, University of California, 1973.
15. H. Degn, I. Balslev, and R. Brook (Eds.), *Measurement of Oxygen*, Elsevier Scientific Publishing Company, Amsterdam, 1976
16. I. Fatt, *Polarographic Oxygen Sensors*, CRC Press, Cleveland, 1976.

5

MEMBRANE-COVERED POLAROGRAPHIC DETECTORS—PRACTICAL CONSIDERATIONS

"A thing may look specious in theory, and yet be ruinous in practice; a thing may look evil in theory, and yet be in practice excellent."

Edmund Burke, 1788

Having dealt with the quantitative expressions for the current we can now consider the influence of materials used in a MPOD on the current and also look at the operational characteristics of a detector. Initially the discussion will be restricted to the case when the steady-state current of a detector is being monitored.

5.1. DESIGN FACTORS

5.1.1. MEMBRANE

The unique characteristic of the MPOD is the membrane which, as noted earlier, serves both to protect the sensing electrode and to give a reproducible electrolysis solution. The theory shows that in addition it provides a finite diffusion layer of controlled thickness at the electrode, and eq. (4.17) for the steady-state current confirms one's intuitive feeling that the more permeable and the thinner the membrane is the greater will be the sensitivity of the detector. Sawyer and co-workers (1) have made a systematic study of some of the membrane materials that are commonly used for MPODs— Mylar, Teflon, polyethylene, natural rubber, silicone rubber, and PVC— while Hoare (2, Ch. VI) mentions a number of additional membranes that have been used; for example, cellophane.

The permeation of polymers to small gaseous molecules depends, as has been seen [eq. (4.11)], on the solubility and diffusivity of the gas in question. For both these quantities reasonable estimates are possible if some basic data on the permeating molecules (e.g. critical temperature and diameter) and of the polymer (e.g. structure, glass transition temperature, and crystallinity) are available, and van Krevelen (3, Ch. 18) has given a very useful general account of how to calculate permeabilities in this way [Appendix I].

71

TABLE 5.1 Comparison of Calculated and Experimental Values of Permeability Data for Oxygen in a PTFE Membrane

	$D_m \times 10^7$ (cm^2 s^{-1})	$P_m \times 10^7$ (cm^2 s^{-1})	K_b
Calculated	1.0	13.7	13.7
Experimental	1.1	8.2	7.5

The values calculated can be compared with experimental data obtained from measurements with a MPOD—Table 5.1; the agreement is fair. The method by which D_m is obtained will be described later [Section 5.2.4]. P_m is readily found using eq. (4.17), and K_b from the ratio P_m/D_m. For a detector with a PTFE membrane of average thickness of 2.54×10^{-3} cm and an electrode area of 0.31 cm^2 the steady-state current in air-saturated water at 25°C is ~ 10 μA. Assuming a four-electron reduction, this gives $P_m \sim 8.2 \times 10^{-7}$ cm^2 s^{-1} [cf. Appendix H]. Figure I.1, together with the equations in Appendix I, is thus quite useful in evaluating a membrane for use with a MPOD. However, although on this occasion the estimate agrees fairly well with what is found in practice some caution is needed in general. This arises from the fact that the estimate is based on diffusion across a membrane in the gas phase; that is, there is gas on both sides of the membrane. With a MPOD used for measuring oxygen dissolved in water the membrane is in contact with a liquid on both sides. Table 5.2, taken from the work of Yasuda and Stone (4), shows that the permeability of oxygen through membranes is rather higher when there is dissolved gas on both sides of the membrane than when there is just the gas itself on both sides. Similar results were obtained by Peirce (5) who measured the ratio D_{CO_2}/D_{O_2} for various wetted membranes. On the basis of Graham's law one expects a value of 0.85, but, as shown in Fig. 5.1, values greater than unity are found. Various explanations have been given to account for this effect (4–6), but the simplest is that the membrane is wetted and oxygen, which diffuses more rapidly in

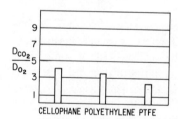

$\dfrac{D_{CO_2}}{D_{O_2}}$

CELLOPHANE POLYETHYLENE PTFE

Fig. 5.1 Comparison of D_{CO_2} and D_{O_2} for various membranes. (Redrawn from reference (5)).

water than in polymer, can permeate not through the membrane itself but through the water existing as molecular clusters in the polymer voids. This idea is supported by the fact that in both Table 5.2 and Fig. 5.1 PTFE, which does not absorb water readily, shows the smallest deviation from the behavior expected for pure gaseous diffusion.

TABLE 5.2   Permeability of Gaseous and Dissolved Oxygen in Polymers (4)

Polymer	Permeability $\times 10^{10}$ $[cm^3 \cdot cm \, cm^{-2} \, s^{-1} \, (cm \, Hg)^{-1}]^a$	
	G/G^b	W/W^c
Poly(dimethylsiloxane)^d	665	4000
Polyethylene—low density	2.34	50.0
Poly(fluorinated ethylenepropylene)—Teflon FEP	3.86	105
Poly(tetrafluoroethylene)—Teflon	23.7	91.0

^{a} The unit is the cgs unit of permeability which expresses the volume of gas at STP that permeates a 1 cm thickness of membrane per square centimetre per second with a pressure drop of 1 cm Hg across the membrane [cf. Appendix A].
^{b} G/G is when there is gas on both sides of the membrane.
^{c} W/W is where there is dissolved gas on each side.
^{d} Dow Corning "Medical Silastic."

Although in principle a wide range of membranes could be used for MPODs, in practice the majority of detectors reported have used either polyethylene or fluorinated plastics (e.g. PTFE or FEP) for a number of reasons. First, they are readily available in thin sheets. Second, they have adequate physical strength. Third, they have, in general, permeability characteristics which do not vary greatly with time, although in the presence of organic vapors significant changes can occur (72). Fourth, they have a low degree of crystallinity and so favor high diffusion rates [cf. Appendix I]. However, the silicone polymer membranes, such as those manufactured by Dow Corning and General Electric, might well be interesting for use with a MPOD (34) because of their even higher oxygen permeability; cf. Table 5.2. The thickness of membranes commonly used at present is usually 10–25 μm, but the use of thinner membranes has also been described. For example, silicone rubber can be stretched over the end of a MPOD so that it is only 1–2 μm thick (70, p. 42), and PTFE membranes only 0.2–1.0 μm thick have been produced by a casting technique (71, p. 86). The main advantage of using these membranes is that the response time to step changes in oxygen is decreased [cf. Section 5.2.4].

The method of fixing the membrane over the electrolyte reservoir and electrodes can be by a variety of mechanical means. Figure 4.1a shows fixing with an O-ring; other fixings such as plastic sleeves (7) and a clamp (8) have been described. Whatever method of fixing is used, three basic requirements are necessary. First, that the membrane should, when in position, be free of creases and folds, particularly over the cathode in order that a uniform surface is available for oxygen transport; some stretching of the membrane usually occurs when it is fixed in position, but this generally becomes reproducible with practice. Second, it is necessary that the membrane once on should not move or vibrate. A movement of the membrane can affect the diffusion process to the electrode (see below), while a vibration of the membrane can produce an alternating component in the cell current which could be rectified and added to the output signal; clearly, a well clamped membrane is necessary to overcome the first problem, and the second problem can be minimized by ensuring that the membrane area is kept small. The third requirement is that the electrolyte between the membrane and the cathode should be a thin, uniform film of a thickness no more than that of the membrane. This need arises from the theoretical considerations that in order to have transport control by the membrane $a/D_e^{1/2} \ll b/D_m^{1/2}$. Since we have seen that $D_m \sim 10^{-2} D_e$ then the inequality only holds if $a \ll 10b$. In practice, a MPOD of the type illustrated in Fig. 4.1 has $a \simeq b$ (9) which means that the inequality is just beginning to hold. There is thus partial joint control of the oxygen transport and this can have a significant effect on the detector characteristics [cf. Appendix H]. In particular it leads to a sluggish response to changes in oxygen concentration; Lucero (10) has given a useful account of this effect.

It is not easy to make the electrolyte layer much thinner than ~ 10–20 μm and so improve the response characteristics, although a special system of threads which allows careful adjustment of the cathode relative to the membrane has been described (71, p. 86). The problem can, however, be obviated if the cathode is in the form of a thin, gas permeable deposit on the inner surface of the membrane; the metal can be put down on the plastic by conventional electroless or vacuum deposition techniques. A composite membrane of this type leads to dramatically improved performance in the time response of a detector and also simplifies the overall design somewhat (11, 12).

5.1.2. ELECTRODES

The oxygen electrode reaction is probably the most complicated one that is commonly encountered. This complexity arises for a number of reasons. First, as was noted previously [Section 3.2], the oxygen reaction is so very irreversible (i.e. the exchange current density is so low) that even minute traces of depolarizing impurities in the solution can affect the overall electrode kinetics and become potential determining. For the overall electrode reaction in acid solution,

$$O_2 + 4H^+ + 4e^- \rightleftharpoons 2H_2O, \qquad E^0 = +1.229 \text{ V}, \qquad (5.1)$$

and in alkaline solution,

$$O_2 + 2H_2O + 4e^- \rightleftharpoons 4OH^-, \qquad E^0 = +0.401 \text{ V} \qquad (5.2)$$

both potentials being with respect to the normal hydrogen electrode. (The significance of the sign of the potentials is discussed in Appendix K.)

But in practice it is very difficult to observe the reversible values, and this is a direct consequence of the low exchange current density and the role of impurities (15, Ch. 10). The current density close to the reversible potential is given by eq. (3.14):

$$i \simeq \frac{i_0 n F \eta}{RT} \qquad (3.14)$$

Then if a deviation from the reversible potential of > 1 mV ($\eta > 1$ mV) is considered significant, with $i_0 \sim 10^{-10}$ A cm^{-2} for the oxygen reaction and with $n \sim 2-4$, impurities giving rise to a current $\sim 10^{-11}$ A cm^{-2} will have an effect on E^0. From eq. (3.30)

$$i = \frac{n F D C_\infty}{\delta}$$

and assuming a diffusion limited current for the impurities, the concentration of impurities affecting the standard potential will be $\sim 10^{-11}$ mol liter^{-1}. It was not until rigorous precautions were taken to exclude low levels of impurities that the equilibrium potential could successfully be reached. In general, however, the open circuit potentials found on noble metals are several hundred millivolts cathodic of the calculated value and depend on, in addition to pretreatment of the solution, electrode history. Bockris and Reddy (15, Ch. 10), Hoare (2, Chs. II and V), and Damjanovic (16) discuss these problems in more detail.

A second reason for the complicated behavior of the oxygen electrode is that the overall reaction

$$O_2 + 4H^+ + 4e^- \rightleftharpoons 2H_2O$$

at low values of pH, or

$$O_2 + 2H_2O + 4e^- \rightleftharpoons 4OH^-$$

at higher values can be regarded, in the simplest terms, as the result of two reactions. For acid solutions, for example,

$$O_2 + 2H^+ + 2e^- \rightleftharpoons H_2O_2$$
$$H_2O_2 + 2H^+ + 2e^- \rightleftharpoons 2H_2O$$

And these reactions themselves can be represented by a network of reactions; for example, for hydrogen peroxide production (17; 18, Ch. 5)

$$
\begin{array}{ccccc}
O_2 & \underset{}{\overset{e^-}{\rightleftharpoons}} & O_2^- & \underset{}{\overset{e^-}{\rightleftharpoons}} & O_2^{2-} \\
H^+ \updownarrow & & H^+ \updownarrow & & H^+ \updownarrow \\
HO_2^+ & \underset{}{\overset{e^-}{\rightleftharpoons}} & HO_2 & \underset{}{\overset{e^-}{\rightleftharpoons}} & HO_2^- \\
H^+ \updownarrow & & H^+ \updownarrow & & H^+ \updownarrow \\
H_2O_2^{2+} & \underset{}{\overset{e^-}{\rightleftharpoons}} & H_2O_2^+ & \underset{}{\overset{e^-}{\rightleftharpoons}} & H_2O_2 \\
\end{array}
$$

In other words, the reaction proceeds by a series of electron and proton transfers, and there are clearly a number of possible routes and sequences by which these can occur. And all this assumes only homogeneous kinetics. If one also recognizes the possibility of surface reactions playing a part, as in fact does occur, then the picture is complicated even further; for example, if we represent a site on the electrode surface by M (15, Ch. 10)

$$O_2 + 2M \rightleftharpoons 2MO$$
$$MO + H^+ + e^- \rightleftharpoons MOH$$
$$MOH + H^+ + e^- \rightleftharpoons M + H_2O$$

Furthermore, in those mechanisms where H_2O_2 can appear as a stable intermediate then it can be reduced on the cathode, or it can decompose heterogeneously on the electrode surface,

$$2H_2O_2 \rightleftharpoons 2H_2O + O_2$$

or it can diffuse away from the electrode into the bulk of the solution. Depending on the relative rates of these processes (16; 19, Ch. 8), four or less electrons will be required for the cathodic consumption of an oxygen molecule. Also, if the H_2O_2 produced is only partially removed by further reduction an appreciable time may be needed to get a steady current at a given cathodic potential while one waits for a steady bulk concentration of the peroxide to build up. The role of H_2O_2 in oxygen reduction has been discussed in general by Breiter (20), Damjanovic (16), and Hoare (2, Ch. IV), and more particularly in the context of MPODs by Hahn, Davis, and Albery (64).

A third complicating factor for the oxygen electrode is, as indicated above, that in the potential range in which oxygen reduction occurs the electrode surface may be covered with oxide and the reaction may involve adsorbed species. This leads to a fourth complication. The adsorbed layers can become so thick and so slow to respond to changes in electrode potential that measurements can easily be carried out at the same potential on surfaces of quite different character. And there are still more complications (e.g. few materials can withstand the highly positive electrode potentials associated with the oxygen reduction without themselves undergoing dissolution, hence contributing to the overall current), but here we have probably given sufficient indication for the justification of our opening statement about the oxygen electrode reaction.

Now, of course, the important question is: "How does all this affect the reduction of oxygen at a MPOD?" Figure 5.2 shows a calculated, steady-state i—V curve for an oxygen detector. The calculation is based on the following equation for the superposition of charge transfer and diffusion control:

$$i = i_0 \left(1 - \frac{i}{i_L} \right) \exp\left(-\frac{\alpha n F \eta}{RT} \right) \tag{5.3}$$

This equation is derived from eqs. (3.12), (3.13), and (3.30). In applying it the assumption has been made that transport through the membrane is rate determining and so the permeability coefficient has been used in place

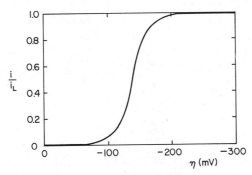

Fig. 5.2 Calculated current-voltage curve for a MPOD. This curve is based on eq. (5.3) where $i_L = nFP_mC_s/b$ and the following values have been used: $i_0 = 10^{-10}A$ cm^{-2}; $\alpha = \frac{1}{2}$; $n = 4$; $P_m = 10^{-6}$ cm^2 s^{-1}; $C_s = 2.5 \times 10^{-4}M$; $b = 2 \times 10^{-3}$ cm.

of the diffusion coefficient. The curve shows a plateau for $\eta > -200$ mV. So if one is working in a very clean, strongly acid solution then a limiting current at $\sim +1.0$ V (vs. NHE) would be expected. Figure 5.3 shows an experimental i—V curve for a MPOD with a gold cathode, and a plateau has formed by ~ 0.3 V from the rest potential. Some cathodic displacement is observed because of the ohmic drop in the thin electrolyte film between the membrane and the cathode. With a solution of, say, $1M$ KCl the ohmic resistance of a typical cell is $> \sim 1$ kΩ and so a current $\sim 3 \times 10^{-5}$ A produces an iR drop of tens of millivolts.

Some comment is necessary about the absolute values of the potentials in Fig. 5.3. The plateau is at potentials cathodic of ~ -0.5 V rather than at $< +1.0$ V for a combination of reasons. The rest potential, as we have noted, is not generally observed at the theoretical value of $+1.23$ V, but rather at $\sim +0.9$ V. And these potentials are for acid solutions. With an unbuffered solution with an initial pH ~ 7, a current of $\sim 3 \times 10^{-5}$ A cm^{-2} flowing for ~ 60 min will produce $\sim 10^{-7}$ moles of OH$^-$ over the electrode area, and since the electrolyte volume in this region is $\sim 10^{-3}$ cm^3 the [OH$^-$] rapidly builds up to ~ 0.1 M or a pH ~ 13 [cf. Appendix J]. With such a high local pH the observed rest potential would be expected to be not at $+0.9$ V but at $\sim +0.1$ V. This is $+0.1$ V with respect to the NHE though, whereas in Fig. 5.3 the potentials are referred to a Ag/AgCl reference electrode which is $\sim +0.22$ V versus NHE. Then on our scale E_e will be ~ -0.1 V, which is close to what is observed, with the plateau several hundred millivolts further cathodic.

Before passing on to discussing specific electrode materials it is worth noting that the relatively anodic value of the equilibrium potential and of the potential for O_2 reduction means that gases, such as SO_2, which are commonly encountered in the environment will not interfere with the oxygen measurements (1). We have only to consider the standard potential

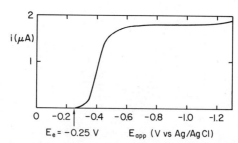

Fig. 5.3 Experimental current-voltage curve for a MPOD. This curve was obtained with an Orbisphere Model 2101 sensor. Cathode—Au; electrolyte—KCl; membrane—Teflon FEP; temperature—25°C.

for the electrode reaction involving SO_2 to see this:

$$H_2SO_3 + 4H^+ + 4e^- \rightleftharpoons S + 3H_2O \qquad E^0 = +0.45 \text{ V}$$

$$\text{cf. } O_2 + 4H^+ + 4e^- \rightleftharpoons 2H_2O \qquad E^0 = +1.229 \text{ V}$$

Which materials can actually be used for the cathode in a MPOD? The following criteria can be listed:

1. High exchange current density.
2. No thick adsorbed layers.
3. No corrosion at the potential for oxygen reduction.

These criteria in fact limit the choice almost completely to the noble metals—Pt, Pd, Rh, Ir, and Au. Even on these metals the oxygen reaction is highly irreversible, but if other materials are chosen—for example, Ni or C—the exchange current is yet still lower (2, Ch. VII; 20). Furthermore, the thick adsorbed layers of oxygen that one finds on many nonnoble metals do not, in general, lead to an increase in the rate of a process occurring by a mechanism of the type involving a metal atom, as given above, but rather, contrary to what one might expect, show the opposite behavior: the net process is inhibited by oxygen layers. Taking the example of Fig. 5.3, of oxygen reduction in essentially an alkaline solution, it is seen that it is necessary to keep the potential constant (potentiostatic condition) at > -600 mV cathodic of the standard potential for the oxygen electrode. Translating this into a potential referred to the NHE gives a value ~ -0.4 V. At this potential many metals form a relatively thick oxide or hydroxide corrosion layer, in contrast to the monolayer for noble metals, and because such an oxide film may act as a nonconducting insulator, an electronically conducting semiconductor, an ionically conducting defect structure, or may simply be reduced, the electrochemical behavior of such a system is complicated and may interfere with the oxygen reduction process. Ag, Pb, and Ni are three such metals, although in the case of Ag the oxide layer is stripped away fairly rapidly (2, Ch. VII), partly because of the high redox potential for the Ag_2O/Ag couple:

$$Ag_2O + H_2O + 2e^- \rightleftharpoons 2Ag + 2OH^- \qquad E^0 = +0.35 \text{ V}$$

$$\text{cf. } PbO + H_2O + 2e^- \rightleftharpoons Pb + 2OH^- \qquad E^0 = -0.58 \text{ V}$$

$$Ni(OH)_2 + 2e^- \rightleftharpoons Ni + 2OH^- \qquad E^0 = -0.72 \text{ V}$$

Once the Ag_2O has been removed oxygen reduction can take place on an oxide-free Ag site.

Of the nonnoble metals therefore, silver is the only real contender as a cathode in a MPOD, but it does suffer from the serious drawback that it is readily poisoned, especially by sulfur impurities [Section 5.3.5]. Various nonmetallic materials have been suggested as cathodes for oxygen reduction, for example, boron carbide (21), nickel arsenide, and nickel sulfide (22), but none seem to offer any real advantages over noble metals. Carbon in various forms has also been examined (2, Ch. VIII; 73, p. 4), and although it does have a number of advantages (e.g. no surface oxides, and ready renewal of the surface by machining) it also has several drawbacks, such as large residual currents and low exchange current.

Of the noble metals Pt and Au have been most widely used in detectors. The other metals in this class do not have any characteristics which would give them a real advantage over Pt or Au, and, in fact, there would be some disadvantages in that Ir and Rh, for example, show problems with thick oxide coatings (16). On both Au and Pt, there is evidence for H_2O_2 production as an intermediate at lower overpotentials (16). On Pt in alkaline solutions the mechanism involves two parallel paths. In one path H_2O_2 is an intermediate which is partially reduced further, but the major reaction path is complete reduction without peroxide formation (16). On Au at high potentials the reduction proceeds along a single reaction path with H_2O_2 as an intermediate which reduces further (16); at low overpotentials some is lost into solution before reduction occurs—the rate constant for diffusion is larger than that for electron transfer—but at high η electroreduction predominates over diffusion. The formation of H_2O_2 will thus probably not be a potential source of complication on either metal especially since the applied potential is very negative and the electrolyte has a high pH in the cathode region. On this basis, therefore, there is not really much to choose between Au and Pt. On other grounds, though, Au has a number of advantages.

Gold only weakly adsorbs oxygen on its surface from solution (2, Ch. VI). Also oxide formation does not begin until a potential more noble than 1.3 V is reached and consequently, while preanodization does not enhance the reversible behavior, in contrast to the case with Pt (73, p. 42), ageing effects which are found in Pt (71, p. 81) are not present on Au. Many workers prefer Au to Pt because of the lack of complications of adsorbed oxygen and oxides (2, Ch. VI).

In conclusion it can be said that Au is probably the best material for the cathode of a MPOD, especially when account is taken of its high resistance to poisoning by noxious gases, notably H_2S [Section 5.3.5]; on this count Ag undoubtedly shows the poorest performance.

The cathode area is usually chosen to give a total current that is suitable for amplification; typically $i_L \sim 3 \times 10^{-5}$ A cm^{-2}, so the required area is readily calculated. Obviously, the larger the current the less the problems of signal handling, but against this has to be balanced the detector wearout characteristics resulting from consumption of the anode and of the electrolyte solution.

All that has been said thus far on electrodes has concerned the cathode reaction and the cathode material. We now briefly turn our attention to the anode. In a voltammetric device the anode must serve two functions: it must act as an auxiliary electrode and as a reference electrode. The reference couples commonly used are Ag/AgCl ($+0.222$ V vs. NHE at $25°C$) or Ag/Ag$_2$O ($+0.35$ V vs. NHE). These electrodes function by the Ag being oxidized in the presence of Cl$^-$ or OH$^-$ ions, respectively:

$$Ag + Cl^- \rightarrow AgCl + e^- \tag{5.4}$$

$$2Ag + 2OH^- \rightarrow Ag_2O + H_2O + 2e^- \tag{5.5}$$

In order to function well as a reference electrode a couple must exhibit a stable potential so that the reference level does not change. The Ag/AgCl and Ag/Ag$_2$O systems form uniform, adherent deposits on the silver electrode and are reproducible and reliable. There are many other couples which could be used instead, but none are really as convenient or as practical as the two mentioned here.

Another requirement of the reference-auxiliary electrode is that it should have sufficient mass in order that its consumption is negligible during the lifetime of the detector. The wearout time, t_{wt}, is expressed by the relationship [cf. eq. (3.1)]

$$t_{wt} = \frac{nFN}{i_{Av}} \tag{5.6}$$

where N is the number of moles of the reference electrode which must be consumed for wearout, and i_{Av} is the average current over the time t_{wt}. One gram of silver in a detector with $i_{Av} \sim 10^{-5}$ A will have a wearout time of nearly three years of continual use, and so electrode consumption should not generally be a problem.

It is also important that the reference electrode surface area be large enough so that polarization at this surface does not become a potential controlling factor. Again this is not usually a problem for the current density at the anode is usually much less than the limiting current density where concentration polarization sets in. Taking the case of the Ag/AgCl couple in $1M$ KCl electrolyte solution, and making use of eq. (3.30) with $\delta \sim 10^{-2}$ cm for a stationary electrode, we calculate $i_L \sim 10^{-1}$ A cm^{-2}. With $i \sim 10^{-5}$ A cm^{-2} a reference electrode of 1 cm^2 area is more than adequate to prevent concentration polarization becoming a problem.

In a galvanometric detector the other electrode is such that its electro-chemical reaction when coupled to that of the oxygen reaction gives rise to a spontaneous current [Appendix K] and the detector functions essentially as a battery. Two systems that are commonly used are the $Cd/Cd(OH)_2$ couple and the Pb/PbO couple:

$$Cd(OH)_2 + 2e^- \rightleftharpoons Cd + 2OH^- \qquad E^0 = -0.761 \text{ V}$$
$$PbO + H_2O + 2e^- \rightleftharpoons Pb + 2OH^- \qquad E^0 = -0.576 \text{ V}$$

Combining either of these electrode reactions with that for oxygen gives the overall cell reaction:

$$O_2 + 2H_2O + 4e^- \rightleftharpoons 4OH^- \qquad\qquad E^0 = +0.401 \text{ V}$$
$$2M + 4OH^- \rightleftharpoons 2M(OH)_2 + 4e^- \qquad E^0_{Cd} = -0.761 \text{ V}$$

$$\overline{O_2 + 2M + 2H_2O \rightleftharpoons 2M(OH)_2} \qquad\qquad \begin{array}{l} E^0_{Pb} = -0.576 \text{ V} \\ E^0{}_{Cell,Cd} = +1.162 \text{ V} \\ E^0{}_{Cell,Pb} = +0.977 \text{ V} \end{array}$$

Similar considerations to those applied to the voltammetric detectors for electrode stability and wearout need also to be applied here. Other couples for galvanometric detectors that have been used are $Pb/PbSO_4$ ($E^0 = -0.356$ V) (7) and Pb/PbI_2 ($E^0 = -0.358$ V) (23). Hoare (2, Ch. VI) has discussed oxygen detectors with Cd and Pb anodes.

5.1.3. ELECTROLYTE

The electrolyte for a MPOD has to be chosen with the anode reaction mainly in mind, but this still leaves plenty of scope and detectors have been described with solutions over the whole pH range and with various cations and anions. Mancy (7) examined acid, neutral, and alkaline solutions and found that although all functioned satisfactorily in a galvanic detector with a Pb anode, the alkaline solution had the lowest residual current. An alkaline solution has a further advantage since, as the oxygen reduction leads to an increase in pH, the OH^- ions can be removed as they are formed by precipitation at the anode. On the other hand, a problem with strongly alkaline electrolytes is the interference by carbon dioxide (11); to lower the effects of this saturated solutions of $KHCO_3$ have been used (24, 25). Electrolytes for use at high and low temperatures have also been reported (2, Ch. VI; 23, 68).

As far as the conditions for oxygen reduction at the cathode are concerned the choice of electrolyte is not so critical since, as noted earlier, there will be a dramatic increase in pH in the vicinity of the cathode [Appendix J], and this means that, apart from an initial short period when the detector is first used, the reduction will always be occurring in an alkaline medium. However, clearly it will take a considerable time for the whole electrolyte volume to follow suit, especially if the electrolyte is buffered (26). Nevertheless, if a chloride containing electrolyte is used then the $[Cl^-]$ will fall as it is consumed by the anode reaction and will be replaced by OH^- generated by the oxygen reduction. Consequently the reference couple will gradually change from $Ag/AgCl$ to Ag/Ag_2O, if Ag is being used as the anode material. This need not present too great a problem though, for provided that the current plateau for oxygen reduction is sufficiently wide the difference in potential between the two reference couples (~ 120 mV) will not be significant.

Another effect of change in electrolyte conditions with the running of a MPOD will arise from the fact that the path for reduction can change with pH and require four or less electrons, depending on the mechanism. Therefore in eq. (4.17) n may be varying with time. This effect should not, however, cause too much problem, except in the initial period when the detector is first used and the pH is changing rapidly [Section 5.2.4 and Appendix J].

A rather serious problem associated with the electrolyte solution is the loss of solvent, almost without exception water, by evaporation and diffusion through the membrane. This solvent depletion is particularly acute when the detector is used for measuring oxygen concentrations in a gaseous phase. Lucero (10) gives an expression for the time required for complete loss of solvent by evaporation, but failure of a detector may well occur long before complete depletion owing to a large increase in the ohmic resistance of the cell as breaks in the conducting path between the electrodes appear.

Several methods can be used to minimize the solvent loss. One method is to have a large electrolyte reservoir (23) and means of ensuring good solvent transfer from the reservoir to the electrolyte film where, of course, the evaporation loss is most serious; capillary pumping action and a small pressure head (27) are two means which have been suggested to facilitate this transport (10). Another method is to reduce the vapor pressure of the solvent by having the electrolyte in the form of a gel or paste instead of a liquid. This method has the additional advantage that as the solvent is depleted, albeit more slowly than from a normal liquid electrolyte, the gelling agent dries into a porous or skeletal structure which provides a suitable medium for capillary pumping. A third method is to add a small amount of deliquescent salt to the electrolyte. Provided that the detector is not in a completely dry atmosphere this can prolong its life considerably—Table 5.3. Finally, a very simple expedient whenever a MPOD is not in continuous use is just to place the detector in a moist or wet environment after each measurement. All this has implicitly assumed that the solvent used is water. Reduction of oxygen in nonaqueous solution has been studied (2, Ch. VI), but to our knowledge no use has been made of such solvents in MPODs. If for any reason a nonaqueous solvent were used though, all of the above methods for reduction of solvent loss, apart from that using deliquescent salts, could be employed.

TABLE 5.3 Comparison of Solvent Evaporation with and without a
Deliquescent Electrolyte

	Electrolyte A (without deliquescent salt)[1]		Electrolyte B (with deliquescent salt)[1,2]	
Time (days)	Weight of solution (g)	Loss of weight[3] (%)	Weight of solution (g)	Loss of weight[3] (%)
0	0.5675	0	0.8499	0
8	0.4817	15.1	0.7448	12.4
24	0.4242	25.3	0.7486	11.9
190	0.3391	40.3	0.7671	9.7

Notes: 1. Electrolyte A: $2.33M$ KCl (half saturated).
Electrolyte B: $2.33M$ KCl + $0.01M$ KH_2PO_4.
2. KH_2PO_4 is a deliquescent salt.
3. Since about 90% of the weight of the solutions is due to the water, the percentage loss of weight is approximately equal to the percentage loss of the total water.

Minimization of solvent loss by evaporation would suggest the need for a large electrolyte reservoir. On the other hand, using a large volume of electrolyte has the serious drawback that in samples with low oxygen concentration oxygen already dissolved in the bulk of electrolyte can contribute significantly to the residual current (7). If it is found to be absolutely necessary to have a large electrolyte volume then this effect on the current can be reduced by making the path between the reservoir and the electrolyte film as long and as tortuous as possible.

The electrolyte concentration used is largely determined by the need for a good conducting medium between anode and cathode; solutions $1M$ or greater are adequate in general. The cations must, of course, be electrochemically inactive at the potential where oxygen reduction occurs, and alkali metal salts are commonly used. The anion, as we have seen, is determined by the reference couple chosen.

A recent and very useful development has been the use of a solid electrolyte in place of the conventional liquid form that we have discussed so far. Niedrach and Stoddard (28) have described a sensor with an ion exchange membrane as electrolyte. An ion exchange membrane consists of an ionic material—usually sulphonic acids for cation membranes and quaternary ammonium compounds for anion membranes—homogeneously incorporated into a polymer matrix, such as polystyrene or halogenated hydrocarbons. The membranes are similar in appearance to any normal plastic sheet, although they are generally thicker (~ 0.1 mm or more) and often have a yellow/brown color.

The mechanism of ion transport through the membrane can be thought of very simply as a hopping process. For an anion exchange membrane:

Anode, e.g., Ag

Porous cathode, e.g., Au

R′ represents the polymer backbone of the membrane and each OH^- is weakly bonded to a quaternary ammonium group which is covalently held to the polymer chain. The hopping occurs under the influence of the electric field across the membrane, and clearly for efficient transport the two electrodes should be in intimate contact with the surfaces of the membrane; this can be achieved by deposition, either electrolytically or under vacuum. At the cathode oxygen is reduced in the normal way,

$$O_2 + 2H_2O + 4e^- \rightarrow 4OH^-$$

with the water being provided from the surroundings and the membrane, which normally is in a wetted state. The cathode needs to be porous in order that the oxygen can diffuse through it and be reduced at the cathode—membrane interface. A normal gas-permeable membrane in addition is not essential for the operation of the device, although since the purpose of MPODs is to protect the cathode against poisoning one would usually use such a membrane. To ensure good contact between the gas permeable membrane, the cathode, and the ion exchange membrane the three have to be bonded together. Clearly an oxygen detector with an ion exchange membrane as electrolyte offers a number of attractive features, not the least being that it is almost a dry system and so lends itself to a rugged construction and to miniaturization. However, precautions still have to be taken against complete drying out and also interference by carbon dioxide absorption. Further details about ion exchange membranes can be found in reference (29) and their use in fuel cells, which is essentially what we have described here, in references 30–32.

5.1.4. GEOMETRY

Apart from the restrictions on the geometry of a MPOD that are predetermined by operational characteristics such as we have already mentioned (e.g. membrane and electrolyte film thickness, anode mass and surface area, cathode area, and electrolyte volume), there are practically no other design factors which are essential for a sensor. Thus in the literature is found a wide variety of geometries, sizes, and arrangements, determined largely by environmental constraints such as operating temperature limits (33), resistance to shock and vibration, humidity, corrosion, external turbulence, and gas and liquid flow requirements, as well as utility considerations such as maintenance, dimensions, weight, and readout and alarm capabilities (10). To describe all these variations would be tedious and so we mention here only a few of the more interesting features. More general information can be obtained from the references listed in Spiehler's comprehensive bibliography of application areas for oxygen analyzers (34), and for features pertinent to medical applications from references 70, 71, and 73.

In discussing the reference electrode of a MPOD we mentioned that it is important that a uniform, adherent film is formed on the anode during electrolysis. This is necessary for two main reasons. First, that the reference potential remains stable, and second, that the anodic deposit does not flake off and fill the electrolyte reservoir or settle on the walls of the reservoir. The danger of a deposit flaking off is that it can readily form a continuous conducting path between anode and cathode and so lead to a short circuit. Molloy (35) has designed a detector with a grooved annulus in the insulating material surrounding the cathode where he claims loose anodic material gets trapped and is thus prevented from reaching the cathode. It is also interesting to note that Molloy's design has the cathode as an annulus surrounding the anode which is a disc at the bottom of the electrolyte reservoir in the center. Neville (8) has described a sensor which is able to withstand severe shock, shaking, and vibration without affecting its measuring performance. This robustness is achieved by having the membrane clamped to the body of the detector. Stack (23) also claims some resistance to mechanical shock by using a convex cathode over which the membrane is stretched. In this way the "drum" effect of a vibrating membrane caused by external turbulences, and which gives rise to oscillations in the readout, is eliminated. Stack further suggests the use of potassium iodide as an electrolyte because its better solubility at $0°C$. The electrolyte remains in solution at this temperature and the gas permeable membrane is not damaged by the growth of salt crystals. Other workers (33) have used a water—methanol—potassium chloride mixture for low-temperature work.

For many biological or medical applications it is necessary that MPODs are able to be sterilized, and a number of groups of workers (36; 73, p. 37) have described such probes. Typical is the probe of Borkowski and Johnson (68) who have made the sensor body of glass and have used an electrolyte with a high boiling point. Glass is obviously not the most practicable material for a detector body and plastic sensors which can be sterilized have also been reported; for example, Stack's electrode in PVC (23).

For medical applications other special features are also often needed, and two designs are particularly worth mentioning: first, multicathode surfaces for measuring oxygen concentrations at neighboring points on a tissue surface (70, p. 147, 161; 74), and second, needle electrodes of diameters as small as 0.2 μm which also allow a high spatial resolution (71, p. 130; 73, Chs. 5 and 9). One rather interesting method of making multicathode surfaces which has been described (73, p. 33) is to use the standard integrated circuit production method of evaporation, photography, and etching to lay down 161 gold cathodes, each 7 μm in diameter and spaced 60 μm apart, to give a total sensor diameter of about 0.6 mm; the cathodes are covered with a plastic membrane and show characteristics typical of more normal sized electrodes. Fatt (73) has described other detector designs for use in specialized medical environments, and Spiehler (34) has listed references to designs used in other specific types of environment; for example, fermentation tanks and bacterial cultures.

5.2. PRIMARY OPERATING CHARACTERISTICS

The operating characteristics of any measurement system may conveniently be divided into two classes (37, Ch. II). Those characteristics which are dependent on the inherent properties of the sensor are called *primary detector characteristics*, and they include sensitivity, linearity, accuracy and precision, selectivity, time response, and inherent long-term stability. *Secondary detector characteristics* are those which refer to the effect of environmental changes on the system. Examples are the effect of flow of the medium, temperature, ionic composition, pollutants, and so on.

Primary and secondary characteristics may vary for detectors of different types and even for sensors of the same type. Thus it is advisable to establish the characteristics of a given system initially and then to check them at regular intervals.

5.2.1. SENSITIVITY

The sensitivity is usually defined as the smallest change in the measured variable that causes a detectable change in the detector signal. It obviously defines the lower limit of detection, for the greater the sensitivity the better will be the capability of the sensor to distinguish between a signal and the background noise. For MPODs we have been using throughout a current $\sim 3 \times 10^{-5}$ A cm^{-2} for an air-saturated solution in water, corresponding to ~ 8 ppm of O_2 at normal temperatures and pressures. The sensitivity of such a detector is clearly $\sim 4 \times 10^{-6}$ A cm^{-2} (mg liter^{-1})$^{-1}$. This is in the same region as most of the detectors described in the literature (37, Ch. XIII), and indeed this is not surprising since all must show a relationship of the form of eq. (4.17) and the terms in this equation are largely predetermined.

At zero oxygen concentration a residual current is always present in a MPOD. This current arises from a number of sources, the most common being the reduction of some impurity or even of the solvent, particularly if too high a polarizing voltage is used. Another possibility is the reduction of oxygen diffusing to the cathode from the electrolyte reservoir, although this contribution should eventually disappear as the reservoir is flushed out with time. Oxygen dissolved in the insulating body of the sensor, especially if this is a plastic, may also slowly out-diffuse when the sensor is in an oxygen-free environment and so contribute to the residual. Ways of reducing the residual include using a cathode of large area, a small exposure area of the electrolyte reservoir, and a minimum of plastic material in contact with the reservoir. Mancy (7) found under the best conditions a residual 0.1–0.2 μA corresponding to 0.03–0.06 mg liter^{-1} of dissolved oxygen. Others (23, 38) have claimed lower residuals, with Koch and Kruuv (38) extending the measurement range to less than about 2 μg liter^{-1}.

Even if the background current is comparatively large, provided that it is constant, very low oxygen concentrations can still be measured by using an imposed opposing current in the measuring circuit to compensate for the residual. In this way changes in O_2 concentration as low as 0.1 μg liter^{-1} can be detected (39). For normal applications, however, a sensitivity of 4×10^{-6} A cm^{-2} (mg liter^{-1})$^{-1}$, with changes ~ 0.05 mg liter^{-1} being detectable and the lower limit also ~ 0.05 mg liter^{-1}, should be adequate.

5.2.2. LINEARITY OF RESPONSE

A typical oxygen calibration curve at constant temperature is shown in Fig. 5.4 and the response is linear over nearly four orders of magnitude. The method of calibrating a MPOD can take several forms. One method is to use gas mixtures of controlled composition; such mixtures are commercially available or may be made up with calibrated flow meters, as was done in obtaining the data for Fig. 5.4. Zero oxygen concentration is obtained by using a pure inert gas, such as He, Ar, or N_2. This method suffers from the drawback that flushing a solution with a gas is a lengthy procedure in general, although the rotated cell designed by Clem (40) overcomes this. Precautions also have to be taken against back diffusion of oxygen if the experiment is done under ambient conditions. Another method is to measure the dissolved oxygen by a standard, well recognized procedure, such as the modified Winkler method [Section 8.1.1]. With this method, though,

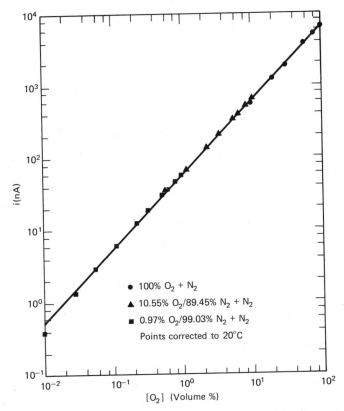

Fig. 5.4 Calibration curve for a MPOD. The different oxygen concentrations were obtained with the gas mixtures shown.

care must be taken since a MPOD measures an intensive property of the solution while a titration, such as the Winkler method, measures an extensive property [Appendix C]. When this precaution is observed good agreement between the two methods is found (41) [Section 5.2.3]. Electrolytic generation of oxygen (65) and the catalase catalyzed decomposition of hydrogen peroxide (66) have also been suggested for calibration of oxygen detectors.

Once a calibration curve has been obtained occasional two point checks should be done with, for example, air-saturated water or air, and oxygen-free water. The aqueous standard sample consists simply of a stirred, aerated volume of clean water at a controlled temperature. The use of air as a calibration standard is advantageous from a number of points of view. First, it is convenient not to have to have a source of clean, aerated water on hand, especially in the field. Second, no stirring of the sample is needed [cf. Section 5.3.1] since diffusion in the gaseous phase is a much faster process than that in water and so no depletion of oxygen in the sample occurs. On the other hand, unless the air is saturated with water vapor the partial pressure of oxygen will not be exactly the same as in the aqueous phase, and an erroneous calibration will be made. Thus in order to use air as a standard great care must be taken to ensure that the air is saturated with water vapor. Holding a detector so that the membrane surface is just above (<1 cm) the surface of a nonpolluted source of water should allow an accurate calibration to be made. A solution of $\sim 2\%$ sodium sulfite provides a very convenient method of obtaining a zero oxygen concentration. The linearity of response shown in Fig. 5.4 is probably adequate for most analyses, especially if one makes use of simple statistics to obtain the least mean squares line and the error (37, Ch. XIV).

5.2.3. ACCURACY AND PRECISION

Accuracy refers to the closeness of a given measurement to the "actual" or "real" value of the quantity measured. Sources of deviation from this "true" value may be found, in which case the readings can be corrected for it. If the deviations arise from indeterminate or random errors statistical treatment of the data gives a measure of this random error. Precision is used to indicate the closeness with which measurements agree with one another, quite independently of any systematic errors involved. An important distinction between precision and accuracy is that accurate measurements are always precise, but the converse is not necessarily true.

Considering the importance of these two characteristics to any analytical technique it is rather surprising to find that so little attention has been paid to reporting measures of the accuracy and precision of MPOD results. Where mention is made of precision it seems generally to be of the order of 0.1–0.2 ppm (26), or a little better under favorable conditions. Table 5.4 shows some data, taken from the work of Reynolds (41), for dissolved oxygen measurements made over a period of time on standard samples using both an oxygen detector and the Winkler method [Section 8.1.1]; the data have been treated statistically. The two methods agree very well and show a precision better than 0.2 mg liter^{-1}; for the Winkler method this is slightly worse than the precision suggested for a modified Winkler method in the Standard Methods of the American Public Health Association (42, p. 410). Claims to much higher precision than ~ 0.1–0.2 mg liter^{-1} for regular measurements are to be taken *cum salis granis*. Other workers (60–62) have also reported good agreement between the Winkler method and measurements with MPODs.

TABLE 5.4 Comparison of
Dissolved Oxygen
Concentrations Measured with a
MPOD and by the Winkler
Method (41)
(O_2 Concentration in mg liter^{-1})

Oxygen sensor	Winkler method
7.8	8.0
8.0	8.2
7.8	7.9
8.3	8.3
8.2	8.3
8.4	8.5
8.5	8.4
7.9	7.7
8.4	8.3
8.3	8.4
8.7	8.6
8.2	8.0
8.2	8.4
Mean 8.21 ± 0.16	8.23 ± 0.16

Note: The confidence limits are cal-
culated from $t\alpha$, where α is the standard
error of the mean and t is the t-table value
at the 95% confidence level.

No systematic study of the accuracy of dissolved oxygen measurements
has been found in the literature, possibly because of the difficulty of obtaining
a standard against which measurements can be made. However, provided
certain precautions are taken in the Winkler method, or in the modifications
of the method [Section 8.1.1], then there seems to be no reason to assume
any systematic error in the measurements, and so the accuracy is likely to
be of the same order as the precision both for the titrimetric methods and
for the oxygen detector. It should be remembered, though, that all this
applies for measurements in the range ~0.1 to ~40 ppm and assumes there
are no complications arising from salting out, for example [Section 5.3.3].

It is worth noting at this point that the results of Table 5.4 show that a
MPOD is as accurate and precise as a standard titrimetric analysis for
oxygen, and in addition, as Reynolds (41) points out, use of an oxygen detector
is easier, less tedious, and considerably faster than wet methods.

5.2.4. TIME RESPONSE

In using a MPOD there are two nonsteady-state situations that are normally encountered. One arises when the detector is switched on, the other when there is an external change in the oxygen level. The first case we have already considered in some detail [Section 4.2], and saw that in theory the steady-state current should be reached after about half a minute from switching on the detector [cf. Fig. 4.5]. In practice the transient current is usually somewhat longer than this—Fig. 5.5. It is unlikely that this is due to delays in the setting up of the steady-state concentration profiles since these are diffusion controlled and have a low energy of activation [Section 3.3.3]. The sluggish response is more likely to be caused by other transient processes occurring at the electrode surface. One such process might be the reduction of a surface oxide on the metal. An oxide film only 10 Å thick would require (if we assume a two-electron transfer, a molecular weight ~ 200, and a density ~ 10) about 10^{-3} C [eq. (3.1)]. With a current density comparable to that for oxygen reduction it would require ~ 100 s to remove the film and for the overall current to reach a steady state.

Another process which could slow the settling down of a detector is the slow adsorption or desorption of impurities in solution. These could cause a change in the charge structure close to the electrode and give rise to a charging current for the resulting varying pseudocapacitance (15, Ch. 9). A third possibility is the complication arising from H_2O_2 being an intermediate in the oxygen reduction [Section 5.1.2]. The most likely transient process to affect the settling down of a detector when it is first switched on is, however, the changing pH in the electrolyte film over the cathode [Appendix J]. This will either cause a change in n in eq. (4.17) or, if the electrolyte film thickness is having a effect on the size of the current being

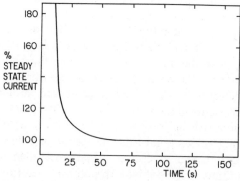

Fig. 5.5 Current transient on switching on a MPOD.

measured [Appendix H], a change in the diffusion conditions close to the electrode.

The second case of a transient current, the response to a step change in oxygen concentration, has not, to our knowledge, been theoretically dealt with for the specific case of a MPOD. Solutions are available, however, for the variation of concentration gradient with time for the simple case of a gas diffusing through a membrane with a step change in concentration at one face of the membrane (43, Ch. 4). This approximates to the case of a MPOD if we consider only transport in the membrane as being rate determining, which will be true under the same conditions as for the case of a potential step.

Figure 5.6 shows transient current curves for a detector with a PTFE membrane on going from deaerated water to air-saturated water and back again (9). In both cases the current is within a few percent of the steady-state value in under 1 min. The transient obtained on going from air-saturated water to deaerated water is slightly more sluggish than the reverse transient and this arises because of nonideal behavior, as discussed later. Other workers (44) have found similar results. That this time scale is reasonable can be shown by using eq. (3.36). Since mass transport by diffusion through the membrane is the slowest step in the overall process, then the response time will be approximately given by

$$t \sim \frac{b^2}{D_m}$$

With $b \sim 20$ μm and $D_m \sim 10^{-7}$ cm^2 s^{-1}, t will be about 40 s, which is in the region that we observe. Commonly 90% of full scale response is found

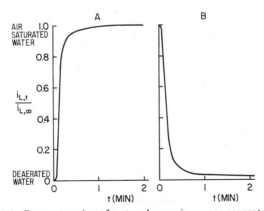

Fig. 5.6 Current transients for step changes in oxygen concentration.

around 20 s and 99% in 40–50 s. Since the response time is proportional to the square of the membrane thickness it can be dramatically shortened by using thinner membranes. With the very thin (e.g. 0.4 μm) PTFE membranes mentioned earlier [Section 5.1.2], 95% response to step changes in oxygen concentration can be attained in 15–20 ms (71, p. 86); this is again in the region that one would estimate from eq. (3.36).

It should be noted that in Fig. 5.6 the symbols $i_{L,t}$ and $i_{L,\infty}$, which were used previously [Sections 3.6 and 4.2] for current transients obtained on switching the electrode potential, are now used to represent the time-dependent and steady-state limiting currents, respectively, when there is a step change in oxygen concentration. Thus, in general, the symbols are for any transient and steady-state limiting currents, and the context should make it clear which type of current transient is being referred to.

Solution of the problem for a concentration step at one face of the membrane while the concentration at the other face is maintained at zero—Fig. 5.7—shows that the variation of concentration gradient within the membrane with time is given by [Appendix L]

$$\frac{\dfrac{\partial C}{dx} - \dfrac{C_{m,\,t=0}}{b}}{\dfrac{C_{m,\,t>0}}{b} - \dfrac{C_{m,\,t=0}}{b}} = 1 + 2 \sum_{n=1}^{\infty} \left[(-1)^n \exp(-n^2\pi^2\tau) \cos\left(\frac{n\pi x}{b}\right) \right] \quad (5.7)$$

where

$$\tau = \frac{D_m t}{b^2} \quad (5.8)$$

and b is the thickness of the membrane.

At $x = 0$, which approximates to the electrode surface if we consider only transport in the membrane as being rate determining [cf. Section 4.2 and Appendix H], the cosine term in the summation reduces to unity and the first term in the numerator on the left-hand side to the concentration

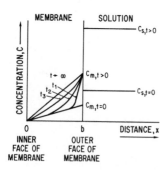

Fig. 5.7 Schematic diagram of concentration profiles within a membrane for a concentration step at one face of the membrane. Subscript "s" stands for "solution", and subscript "m" for "membrane".

gradient at the electrode. The other terms on the left-hand side can also be regarded as concentration gradients at the electrode if we assume linear slopes, and so eq. (5.7) reduces to

$$\frac{i_{L,t} - i_{L,0}}{i_{L,\infty} - i_{L,0}} = 1 + 2 \sum_{n=1}^{\infty} \left[(-1)^n \exp(-n^2\pi^2\tau) \right] = F(\tau) \qquad (5.9)$$

where $F(\tau)$ is just a shorthand way of writing one plus the summation. This equation can be compared to the experimental current—time curve in a number of ways.

One way is to calculate the current ratio from experimental values at regular time intervals, say 2, 4, 6, 8 s, and so forth, from the start of the transient; this gives experimental values of $F(\tau)$. The function $F(\tau)$ can also be computed for values of τ, and the resulting curve used to obtain τ for an experimental value of $F(\tau)$. Since we know the time corresponding to the experimental $F(\tau)$ we can then calculate the value of D_m, the diffusion coefficient of the gas in the membrane, from eq. (5.8). Alternatively we can plot t versus τ for a number of times and therefore obtain D_m from the slope. Figures 5.8 and 5.9 illustrate this procedure. For transient A—de-aerated water to air-saturated water—the t versus τ plot is reasonably linear up to 22 s, but the plot corresponding to transient B—the reverse step—is only linear up to 14 s; the reason for this difference will be discussed shortly. Taking the linear portion of the plot for B and all of the plot for A allows values of D_m to be calculated. The two values obtained—1.17×10^{-7} and 1.11×10^{-7} cm^2 s^{-1}—agree reasonably well, and are also in accord with the value calculated for PTFE using Fig. I.1. The permeability coefficient calculated from the steady-state current is the same as found previously—8.2×10^{-7} cm^2 s^{-1}—and so $K_b \sim 7.5$.

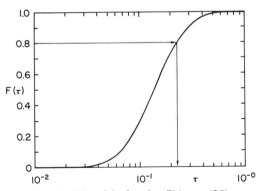

Fig. 5.8 Plot of the function $F(\tau)$—eq. (5.9).

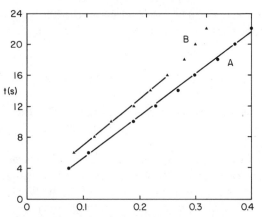

Fig. 5.9   Plot of t versus τ for data taken from Fig. 5.6. An example of a method of obtaining this figure follows: from Fig. 5.6 for transient A, $i_{L,0} = 0$ and at $t = 12$ s, $i_{L,t}/i_{L,\infty} = F(\tau) = 0.8$. From Fig. 5.8, $F(\tau) = 0.8$ is equivalent to $\tau = 0.23$. Above we plot $t = 12$ s and $\tau = 0.23$. The least mean squares slope of the t versus τ plot for transient A is 54.9. From eq. (5.8), with $b = 2.54 \times 10^{-3}$ cm, we calculate $D_m = 1.17 \times 10^{-7}$ cm^2 s^{-1}. For transient B, $D_m = 1.11 \times 10^{-7}$ cm^2 s^{-1}.

Our treatment of the results can be simplified a little if we only examine the current transient at $> \sim 1$ s after its start. With t of this order the argument of the exponential term in eq. (5.9) is large enough so that only the first term in the summation need be considered and the equation reduces to

$$\frac{i_{L,t} - i_{L,0}}{i_{L,\infty} - i_{L,0}} = 1 - 2 \exp(-\pi^2 \tau) \tag{5.10}$$

or

$$\ln(i_{L,\infty} - i_{L,t}) = \ln 2(i_{L,\infty} - i_{L,0}) - \pi^2 \tau$$

$$= \ln 2(i_{L,\infty} - i_{L,0}) - \frac{\pi^2 D_m t}{b^2} \tag{5.11}$$

Figure 5.10 shows a plot of this equation and from the slope of the line we calculate $D_m = 1.14 \times 10^{-7}$ cm^2 s^{-1}, which is in good agreement with the value from eq. (5.9).

A third method of analyzing a current transient of the type in Fig. 5.6 is to plot the integral of the current as a function of time. Integrating eq. (5.9) gives

$$\frac{\int_{t_1}^{t_2} (i_{L,t} - i_{L,0}) \, dt}{i_{L,\infty} - i_{L,0}} = t - \frac{b^2}{6D_m} - \frac{2b^2}{D_m \pi^2} \sum_{n=1}^{\infty} \frac{(-1)^n}{n^2} \exp(-n^2 \pi^2 \tau) \tag{5.12}$$

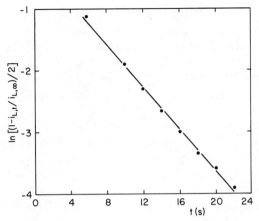

Fig. 5.10 Plot of eq. (5.11) for transient A of Fig. 5.6. The least mean squares slope of the line is -0.174, and taking $b = 2.54 \times 10^{-3}$ cm this gives $D_m = 1.14 \times 10^{-7}$ cm^2 s^{-1}.

which for large values of t simplifies to

$$\frac{\int_{t_1}^{t_2} (i_{L,t} - i_{L,0})\, dt}{i_{L,\infty} - i_{L,0}} \simeq t - \frac{b^2}{6D_m} \qquad (5.13)$$

This equation gives an intercept on the t axis from which the diffusion coefficient can be calculated. Figure 5.11 treats the data in this way and

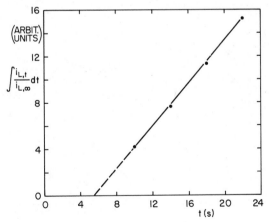

Fig. 5.11 Plot of eq. (5.13) for transient A of Fig. 5.6. When the integral is zero then $t = b^2/6D_m$. From the intercept we calculate $D_m = 1.91 \times 10^{-7}$ cm^2 s^{-1}.

gives D_m in fair agreement with the values found by the other two methods. This method is probably less accurate than the previous two since it depends on the extrapolation of the line to get the intercept, and this is never a very satisfactory procedure. Also the method involves integration of the i—t curve which, unless it is done automatically, is a tedious process. On balance the use of eq. (5.11) is probably to be recommended.

Finally, using an average value of D_m from the first two methods we can plot a current transient according to eq. (5.10) and compare it with the experimental curve. This is done in Fig. 5.12, and there is a good correlation. The value of treating data from a concentration step by any of the methods we have discussed lies not so much in the calculation of the value of the diffusion coefficient, but rather in the fact that it is a good test of whether a MPOD is functioning in an "ideal" fashion. Departures from the behavior expected on the basis of eq. (5.9), (5.11), or (5.13) will indicate that mass transfer is not being determined solely by diffusion in the membrane. Figure 5.9 clearly shows that transient B in Fig. 5.6 departs from ideal behavior. The transient is sluggish since for a given τ value the measured time is too long. Other workers have also reported slow transients (36, 64). It has been suggested (64, 75) that the poor time response on going from an oxygen rich environment to one where there is no oxygen is probably due to the presence of hydrogen peroxide in the electrolyte and the slow rate at which it is consumed at the cathode. A significant improvement in the response to a sudden decrease in oxygen concentration can be achieved by adding a few milligrams of catalase to the electrolyte solution under the membrane; the catalase acts as a catalyst for hydrogen peroxide decomposition. Lucero (10) has given a general qualitative description, in terms of equivalent circuits, of some of the reasons for slow time responses of detectors.

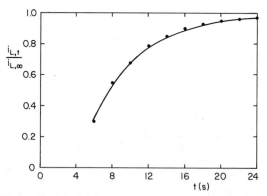

Fig. 5.12 Comparison of calculated transient (continuous line) and experimental transient (points).

5.2.5. STABILITY

The long-term stability of a detector refers to the change in the sensor's performance with time, assuming, of course, that external conditions (for example, temperature) remain constant. We have already indicated a number of causes for changes in the signal output with time: drying out of the electrolyte because of solvent evaporation, unstable reference potential, short circuiting of anode and cathode by loose anode deposit, and localized electrolyte concentration changes. Probably, however, a more important cause of drifting or ultimate failure of a MPOD is contamination or poisoning of the working electrode surface. This can occur by two major mechanisms—deposition of materials from the electrolyte solution, and contamination by gases diffusing through the membrane from the external environment. If it is feared that impurities present in the electrolyte, especially reducible metal ions, are depositing or being absorbed on the cathode surface and are inhibiting its catalytic activity then the solution can be cleaned up beforehand (15, Ch. 10).

Contamination by this means, though, is usually a very slow process and it is possible to operate a detector satisfactorily over reasonable time periods despite the presence of such impurities. The deposition of small particles of anode products on the cathode and their eventual build up to a continuous bridge between the two electrodes is likely to be a more serious limitation on the detector stability. The effect of polluting gases from the environment is essentially a secondary operating characteristic and so this will be dealt with later [Section 5.3.5].

Detectors operating for periods of many months with negligible drift have been reported (25). Figure 5.13 shows the output with time for a MPOD with a Au cathode and Ag anode operating in a clean sample of stirred

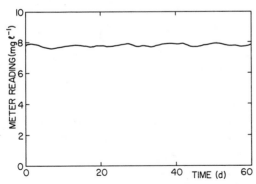

Fig. 5.13 Long-term stability of a MPOD (9). Temperature—25°C; Atmospheric pressure—720 Torr.

water in a thermostat bath. The overall variation in the signal is less than $\pm 2.5\%$, and there were indications that the short term changes were due not to an inherent instability of the detector but to the failure of the thermostat to maintain adequate temperature control.

5.2.6. SELECTIVITY

This refers to the effect of interferences from detectable species other than oxygen. It has already been pointed out [Section 5.1.2] that the relatively anodic value of the equilibrium potential for the oxygen reaction means that gases, such as SO_2, which are commonly encountered in the environment will not interfere with the oxygen measurement. There are, however, other gases that are associated with standard potentials even more anodic than that for oxygen:

$$Cl_2(g) + 2e^- \rightleftharpoons 2Cl^- \qquad E^0 = 1.358 \text{ V}$$
$$2NO + 2H^+ + 2e^- \rightleftharpoons N_2O + H_2O \qquad E^0 = 1.59 \text{ V}$$

In addition, because of the high overpotential needed for oxygen reduction some couples which have $E^0 < 1.229$ nevertheless are active before oxygen; the Br_2/Br^- couple, for example, is very reversible and a diffusion plateau appears for Br_2 reduction many tenths of a volt anodic of the oxygen plateau (1).

$$Br_2(g) + 2e^- \rightleftharpoons 2Br^- \qquad E^0 = 1.087 \text{ V}$$

The problem of interferences of these types is often, however, alleviated somewhat, first by the fact that usually the reducible gases are present only in trace amounts, and second by the membrane acting as a "filter." If a reducible gaseous impurity only contributes at most 1–2% of the detector current then periodic checking and adjustment of the zero current level may well be sufficient to overcome the interference. Even when the background current is a relatively large proportion of the total, provided that it is steady and the interfering gas does not contaminate the system, the detector signal can be compensated for this current. When a large background is not steady, however, the only real solution is to remove the interference either before or during measurement with some gas separation device (10) [Section 8.2].

The membrane acts as a "filter" simply because it shows different permeabilities to different gases. Permeability is a function of the diffusion coefficient and the solubility of the gas in the membrane [eq. (4.11)]. With oxygen analysis there are few, if any, reducible gases likely to be present with a molecular diameter smaller than that of O_2, and, as Fig. I.1 shows, the larger the diameter the smaller the diffusion coefficient for a given membrane. On the other hand, there are also not many gases with as low a critical temperature as that of oxygen and so the solubility of many gases is higher than the oxygen solubility in the same membrane [eq. (I.2)]. Fortunately, however, the diffusion term will generally outweigh the solubility and so the membrane acts in a semipermeable fashion, letting O_2 through more easily than other gases. In general therefore, taking all things into consideration, the selectivity of MPODs is good, and there has been little reported of the type of interference we have mentioned here.

5.3. SECONDARY OPERATING CHARACTERISTICS

5.3.1. FLOW OF THE TEST SOLUTION

In the theoretical derivation of the current for the MPOD it was assumed that the test solution was stirred in order to maintain the bulk concentration, C_s, of oxygen up to the membrane surface. If this condition is not fulfilled then depletion of oxygen in the test solution close to the membrane occurs, the effective diffusion layer thickness is increased, and the current falls. Figure 5.14 shows the effect of stirring on sensitivity.

An estimate of the minimum bulk flow rate of the medium required to prevent excessive depletion outside the membrane can be made by making use of eq. (3.26):

$$\delta_s \sim D_s^{1/3} v_s^{1/6} \left(\frac{x}{v_m}\right)^{1/2} \tag{3.26}$$

where the subscript s refers to the test solution.

With δ_s as the thickness of the diffusion layer in the test medium then we need to know how great the solution flow velocity, v_m, must be in order that transport across this layer does not start to control the current. For diffusion in the membrane to remain rate determining a similar condition to that used in comparing the membrane and the thin electrolyte film must hold,

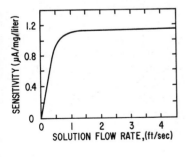

Fig. 5.14 Effect of stirring on sensitivity (37) (Reproduced with permission of Ann Arbor Science Publishers.)

namely

$$\frac{\delta_s}{D_s} \ll \frac{b}{P_m}$$

or

$$\delta_s \ll 10b$$

Taking this condition to hold for $\delta_s \sim 0.5b \sim 10^{-3}$ cm, and evaluating eq. (3.26) with characteristic values for D_s and v, and 0.6 cm for the characteristic length, x, we find $v_m \sim 25$ cm s^{-1}. In Fig. 5.14 the sensitivity has stabilized at a solution flow rate of ~ 1 ft s^{-1} or ~ 25 cm s^{-1}, which agrees with our estimate. For the characteristic values chosen, currents independent of flow rate will not usually be found for $v_m < \sim 25$ cm s^{-1}. This value is based, however, on two assumptions. First, that laminar flow of the test solution prevails. If there are local turbulences close to the electrode a steady current may be obtained at rather lower bulk flow rates. The second assumption is that b, the membrane thickness, is ~ 20 μm and x, the characteristic length (e.g. the cathode diameter), is ~ 0.6 cm. If b is much smaller than 20 μm then to satisfy the inequality $\delta_s \ll 10b$, v_m must be correspondingly greater. On the other hand, we have pointed out [Section 5.2.4] that for fast response to a step change in oxygen concentration b should be as small as possible. A solution to this conflicting requirement can be achieved by reducing the characteristic length, x. So, for example, with a 6 μm PTFE membrane a 95% response to an oxygen concentration change can be obtained in a few seconds, as would be expected from eq. (3.36), but by using a cathode of diameter ~ 1 mm the solution flow velocity requirement can be adjusted so that v_m need only be >5 cm s^{-1} (70, p. 100). Detailed studies of the effect of stirring on sensor current have been made (73, p. 24), and Schuler and Kreuzer (70, p. 64) have examined the interdependence of solution flow velocity and response time, presenting a nomogram for computing the flow requirements in terms of the membrane thickness and cathode diameter. Of course, as the diameter of the cathode gets smaller the total current output falls, and with cathodes $\ll 1$ mm this may give rise to problems of amplification. However, the advantages of both a reasonable cathode area and short characteristic length can be obtained by using an annular cathode with a narrow width.

Adequate mixing of the test solution around the top of the membrane probe can be achieved either by driving the solution past the surface of the analyzer by means of a pump or a stirring device, or by moving the analyzer. In the laboratory a magnetic or motor driven stirrer may be used, while in the field the natural flow of the water, towing the probe behind a boat, or moving the probe up and down manually can provide sufficient agitation. Various attachments using battery driven agitators mounted on the end of the probe have been described (45, 46). If none of these methods is suitable then the effect of stirring can be reduced by increasing the membrane thickness (44), or by having a double membrane system with a thin air gap between the membranes (69); the latter method has the additional feature of showing the same sensitivity in both gaseous and liquid phases. The disadvantage of a thick membrane is the resulting decrease in sensitivity, while both the thick membrane and double membrane systems show a very slow response to changes in oxygen concentration.

5.3.2. TEMPERATURE EFFECT

MPODs show a large dependence on temperature, the signal increasing by 1–6% for a rise of 1°C. Figure 5.15 gives current—voltage curves recorded at various temperatures and without any correction being made, and Fig. 5.16 shows the values of the current at the center of the plateaus plotted as a function of temperature. Also plotted on Fig. 5.16 for comparison is the variation of oxygen solubility with temperature, and clearly measurements with oxygen detectors have to be made under thermostatic conditions or there has to be some method of correcting the observed signal.

The effect of temperature on the process of diffusion of simple gases in membranes is expressed by eq. (3.23)

$$D_m = D_0 \exp\left(-\frac{E_D}{RT}\right) \tag{3.23}$$

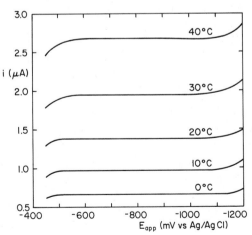

Fig. 5.15 Current-voltage curves at various temperatures (9). These curves were obtained with an Orbisphere Model 2101 sensor. Cathode—Au; anode—Ag; electrolyte—KCl.

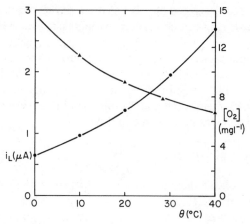

Fig. 5.16 Variation of detector current and oxygen solubility with temperature.

E_D, the activation energy in J mol^{-1}, is, as noted earlier [Section 3.3.3], the energy needed to enable a dissolved molecule to "jump" out of one "hole" or position into another "hole." And since larger "holes" need to be formed in the polymer for the diffusion of larger molecules, the activation energy will be greater for diffusion of bigger molecules and the diffusivity smaller; this is indeed found to be true [cf. Fig. I.1]. The theoretical interpretation of E_D, and also D_0, the preexponential factor, are discussed at greater depth by Kumins and Kwei (47, Ch. 4).

The variation of solubility of a gas in a membrane with temperature can be described by an equation of the same form as that used for gas solubility in a liquid [eq. (2.9)] which, expressed in terms of the distribution coefficient, is written as

$$K_b = K_0 \exp\left(-\frac{\Delta H_m}{RT}\right) \tag{5.14}$$

where K_0 is also a preexponential term and ΔH_m is the heat of solution in J mol^{-1}; van Krevelen (48, Ch. 18) discusses the significance of ΔH_m in more detail.

The permeability coefficient, which is the product of the distribution coefficient and the diffusion coefficient, can thus be expressed by

$$P_m = P_0 \exp\left(-\frac{E_P}{RT}\right) \tag{5.15}$$

P_0 is the product of the two preexponential factors and E_P, the activation energy for permeation, is the sum of the diffusion activation energy and the heat of solution in the membrane. If the approximate expression for oxygen solubility as a function of temperature [eq. (2.9)] is now taken together with eq. (5.15) and they are substituted into the equation for the steady-state current [eq. (4.17)] then

$$i_L \simeq \frac{nF}{b} P_0 \exp\left(-\frac{E_P}{RT}\right) K'_0 \exp\left(-\frac{\Delta H}{RT}\right) \tag{5.16}$$

K'_0 includes the constant term in eq. (2.9) and ΔH is the heat of solution for the gas in the liquid outside the membrane. Combining P_0 and K'_0 in a single constant, P'_0, and E_P and ΔH in a single energy represented by E'_P, eq. (5.16) becomes

$$i_L \simeq \frac{nF}{b} P'_0 \exp\left(-\frac{E'_P}{RT}\right) \tag{5.17}$$

or

$$\ln i_L = -\frac{E'_P}{RT} + \text{constant} \tag{5.18}$$

Figure 5.17 plots the data from Fig. 5.16 according to this equation, and it is seen that there is a reasonable linear relationship between the logarithm of the current and the inverse of the absolute temperature; for convenience $\ln(i_{L,\theta}/i_{L,20})$, where $i_{L,\theta}$ and $i_{L,20}$ are the currents at $\theta°C$ and $20°C$, respectively, has been plotted rather than $\ln i_L$.

It should be noted that this treatment assumes that the only detector parameter of major importance which varies with temperature is the membrane permeability coefficient. Other membrane parameters which could

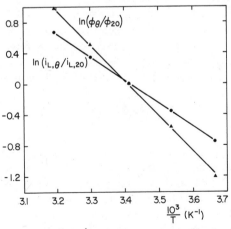

Fig. 5.17 Plots of eqs. (5.18) and (5.19).

contribute to the temperature dependence are the coefficient of thermal expansion, heat distortion, and thermal conductivity, but in general these have a small effect. There are, though, other factors which can cause anomalous temperature variations. For example, many plastic materials undergo a phase change with temperature as a result of changes in the degree of crystallinity. This phenomenon is indicated by a sudden change of slope in the $\ln i_L$ versus $1/T$ plot. Further, expansion and contraction of the electrolyte, and especially of gas bubbles trapped in the thin electrolyte film, can cause dramatic alteration of the signal at various temperatures. Offsets or jogs in the current—temperature plot without a change of slope are characteristic of the failure of the membrane holder assembly to maintain a reasonably constant membrane tension and electrolyte layer thickness as the temperature varies. And even if these problems are not met, often membranes of the same thickness and structural material show different permeability coefficients from one batch to another, depending on the degree of crystallinity of the polymer caused by the manufacturing process [cf. eqs. (I.1) and (I.7)].

In view of all these considerations it is obvious that the temperature response of an oxygen sensor is a critical characteristic of any measurement. Various ways of dealing with the need for temperature correction are available. The simplest method is always to carry out measurements under thermostatically controlled conditions, or to measure the ratio of the signal of the test solution to that of a reference solution at the same temperature. This procedure, however, may not be always practicable or convenient and then a calibration chart is useful. Such a chart may be of the form of the curve in Fig. 5.16 or of the plot in Fig. 5.17. In either case the observed current at a given temperature is compared with the current on the calibration chart at the same temperature, and then, assuming linear response at a fixed temperature, the unknown concentration can be readily calculated.

Another way of plotting the current dependence on temperature is to use the sensitivity coefficient, φ, which is the ratio of the current at a given temperature to the oxygen solubility at that temperature. φ will be related to T by

$$\varphi = \frac{nF}{b} P_0 \exp\left(-\frac{E_P}{RT}\right) \tag{5.19}$$

φ is plotted on Fig. 5.17 together with i_L. This plot can also be used for calibration by determining the sensitivity at a known temperature, plotting the point on Fig. 5.17, and drawing a line through this point and parallel to the line already present. The sensitivity at any other temperature can then be found. The advantage of this method is that it takes care of variations in permeability coefficient. The $\ln \varphi$ plot also has two other advantages. It allows the activation energy of permeation to be calculated. From the slope

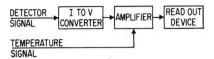

Fig. 5.18 Block diagram of a circuit for a temperature-compensated oxygen detector.

of ln φ versus $1/T$ in Fig. 5.17, E_P is calculated to be 38.9 J mol^{-1}, which is in reasonable agreement with reported values (48, Ch. 18); this method of measuring E_P is a useful and simple alternative to differential pressure methods. The second use of the sensitivity plot is that it simplifies the treatment of automatic temperature compensation.

A perfectly compensated MPOD will have the same sensitivity at all temperatures. If we consider the current signal from the detector being first converted to a voltage and then amplified by a stage which adjusts the amount of amplification according to the temperature (Fig. 5.18), then in order that φ remains constant over the temperature range at which measurements are made (normally 0–50°C) it is necessary to have a temperature element that shows an inverse dependence on temperature. A thermistor, or temperature dependent resistance, exhibits this behavior and its resistance is expressed by

$$R_T = R_0 \exp\left(\frac{A}{T}\right) \tag{5.20}$$

R_0 and A being constants. The significance of these constants is discussed in references 49–51, but we shall not concern ourselves further with it here. If we simply accept that thermistors do have a characteristic behavior represented by eq. (5.20), then knowing the values of the constants the thermistor resistance at any temperature can readily be calculated. Figure 5.19 shows a

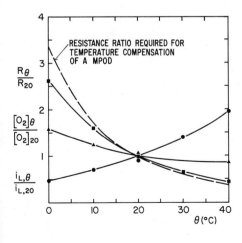

Fig. 5.19 Variation of detector current, oxygen solubility, and thermistor resistance with temperature.

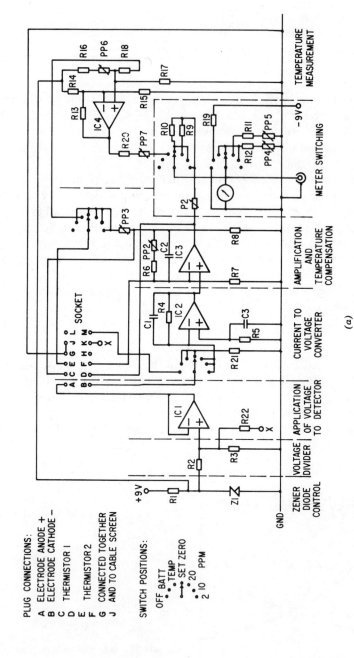

Fig. 5.20 Electronics for an oxygen detector. (a) A typical control circuit. (b) A complete assembly.

(a)

PLUG CONNECTIONS:

A ELECTRODE ANODE +
B ELECTRODE CATHODE−
C THERMISTOR 1
D
E THERMISTOR 2
F
G CONNECTED TOGETHER
J AND TO CABLE SCREEN

SWITCH POSITIONS:

OFF BATT
• • TEMP
• SET ZERO
• 20
2 10 PPM

108

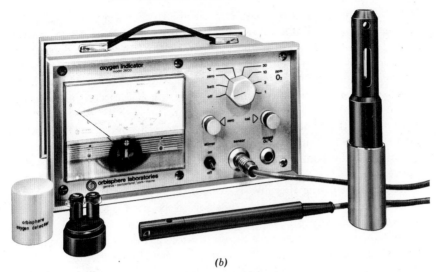

(b)

Fig. 5.20 (cont.)

typical variation of thermistor resistance with temperature; the ratio of thermistor resistance at any temperature to that at a reference temperature has, for convenience, actually been plotted instead of the direct thermistor resistance. Unfortunately, however, thermistors do not show such a large temperature variation as is required—as can be seen from the typical oxygen solubility and current-temperature dependence curves also given in Fig. 5.19—and in order to achieve adequate temperature compensation of the detector signal it is necessary to incorporate two or more thermistors in a circuit network. Briggs and Viney (24) have given an account of various methods of temperature compensation.

One possible circuit is shown in Fig. 5.20a. This circuit operates from positive and negative 9 V lines, supplied either from batteries or from an external supply. A stable voltage of the order of 800 mV is supplied to the oxygen detector by a zener diode stabilized circuit and the potential dividing circuit; operational amplifier IC1, connected as a voltage follower,* has a low-output impedance which ensures stability of the applied voltage against changes of the detector current. The sensor current is fed through the main selector switch to the input of a current-to-voltage converter, IC2, and the output of this amplifier is fed to the amplifying stage, IC3. The feedback circuit of this stage contains two thermistors, which are located close to the

---

* The basic principles of operational amplifiers are covered in references 52–54.

electrode system, and also series and parallel resistors available as preset controls (PP2 and PP3) to allow the slope of the compensation curve to be tailored so that it aligns with that required by the temperature dependence of the detector. The output from IC3 is applied to a calibration control, P2, and to a dc microammeter with appropriate range switching. IC4 with one of the thermistors normally connected into the IC3 circuit, and various series and parallel resistors, allows the temperature also to be measured and displayed on the meter. Figure 5.20b illustrates a complete DO analysis system. Figure 5.21 shows how the circuit compensates the sensitivity data of Fig. 5.17; the compensation is within $\pm 2\%$ over the operating range 0–40°C.

The location of the thermistors should ideally be at the surface of the membrane over the working electrode surface because it is the temperature at this point which will have the greatest effect on the detector signal. It is clearly difficult, though, to place the thermistors at this site without seriously affecting the operation of the detector by obstructing the flow of oxygen through the membrane. Therefore the criterion of a suitable position is where the thermal characteristics match that of the ideal site as closely as possible. Lucero (10) has discussed and analyzed the thermal behavior of various sites, and also concludes that the thermistors should be placed as

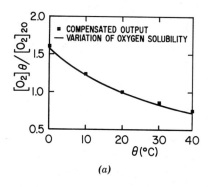

(a)

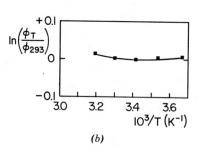

(b)

Fig. 5.21 Temperature compensation of a MPOD using the circuit of Fig. 5.20. (a) Compensated output of a MPOD. (b) Compensation of sensitivity coefficient.

close as possible to both the membrane and the working electrode; one suggestion for doing this is to have a hollow electrode and to place the thermistors inside the electrode.

Automatic temperature compensation obviously offers a number of advantages. But we have mentioned that variations in membrane permeability often arise both during a single series of measurement and from one series to another. Some caution is therefore necessary in using automatic compensation, and certainly frequent checks should be carried out, particularly whenever the detector is subjected to large temperature changes or when the membrane is changed. In addition the detector should always be calibrated at a temperature as close as possible to the sample temperature, for, as Fig. 5.21 shows, the compensation becomes less good the further one moves from the calibration point. For the most accurate work automatic compensation should not be relied on, but one of the other methods of correction suggested above should be used.

5.3.3. SALT EFFECT

In the derivation of the current at a MPOD and in what has followed the concentration of oxygen rather than its chemical potential or activity has been used—it has been assumed that concentration and activity are identical. In test solutions where there is little dissolved salt this is a reasonable assumption since in such solutions the activity coefficient, γ_c, is close to unity [cf. Section 2.4]. On the other hand, in solutions with high salt concentration γ_c is greater than one and the concentration and activity of the dissolved oxygen are no longer equal, the concentration falling and the activity remaining constant. And for MPODs this is an important effect since an oxygen detector responds to the difference in activity across the membrane rather than the concentration difference. So in samples containing electrolyte, while the oxygen concentration falls with increasing salt concentration the detector current remains constant—Fig. 5.22.

Equation (4.17) for the steady-state current density should thus be written

$$i_L = \frac{nF}{b} P_m a = \frac{nF}{b} P_m \gamma_e C_e \qquad (5.21)$$

where γ_e and C_e are, respectively, the activity coefficient and concentration of the dissolved oxygen in a sample containing dissolved electrolyte. In pure water γ_e is unity, $C_e = C_s$ and we have eq. (4.17). The variation of γ_e with dissolved salt concentration can be described by eq. (2.21):

$$\ln \frac{\gamma_e}{\gamma} = k_s I \qquad (2.21)$$

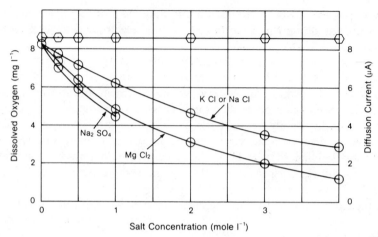

Fig. 5.22 Oxygen concentration and detector current as a function of dissolved salt concentration. Uppermost line is the detector current (7). (Reproduced by permission of Elsevier Sequoia S.A.)

where k_s is a salting-out coefficient (67) and I is the ionic strength of the solution. With $\gamma = 1$ for pure water this equation can be written

$$\gamma_e = \exp(k_s I) \tag{5.22}$$

hence

$$i_L = \frac{nF}{b} P_m C_e \exp(k_s I) \tag{5.23}$$

or

$$\ln \frac{i_L}{C_e} = k_s I + \text{constant} \tag{5.24}$$

The logarithm of the sensitivity coefficient should then be proportional to the ionic strength and Fig. 5.23 shows this to be so. From the gradient of the line the salting-out coefficient can be determined, and from this, using eq. (5.22), the activity coefficient at a given ionic strength; both these quantities are of some theoretical importance (67).

Although it is normal to think in terms of dissolved oxygen concentrations it is actually more significant to define oxygen in solution in terms of activity, since this is the "effective concentration" [Appendix C]. Therefore measurements with a MPOD are usually more meaningful than results from a titrimetric analysis, such as the Winkler method.

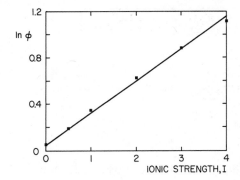

Fig. 5.23 Variation of the sensitivity coefficient with ionic strength—test of eq. (5.24) with data for KCl or NaCl from Fig. 5.22.

When the oxygen concentration is nevertheless required from a measurement by a MPOD in salt solution then eq. (5.23) can be used to calculate it, knowing the salting-out coefficient and the ionic strength; this calculation must, of course, be done using k_s obtained under the same conditions of temperature and pressure. An alternative to calculating the concentration is to have automatic compensation in much the same way that the temperature is compensated. The ionic strength, I, of a test solution can be determined accurately by means of conductance measurements:

$$\varkappa = \xi + \lambda I + \chi I^2 + \omega I^3 \qquad (5.25)$$

where $\varkappa$ is the specific conductance in $ohm^{-1} cm^{-1}$ [Appendix M], and ξ, λ, χ, and ω are all constants. This nonlinear equation approximates to

$$\varkappa = \xi + \lambda I$$

or

$$I = \frac{(\varkappa - \xi)}{\lambda} \qquad (5.26)$$

which holds in most surface waters where the salt content is not too high. The concentration of oxygen is then given by

$$C_e = i_L \frac{b}{nFP_m} \exp\left[-\frac{k_s(\varkappa - \xi)}{\lambda}\right] \qquad (5.27)$$

To convert a current into a concentration requires multiplication by the exponential term, assuming the system is calibrated as usual to take care of the (b/nFP_m) term. Figure 5.24 illustrates the principle of this correction.

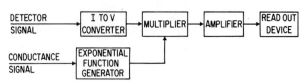

Fig. 5.24   Principle of automatic compensation for the salt effect.

5.3.4.   THE EFFECT OF DEPTH

The opposite effect to that of salting-out will occur when MPODs are used at great depths [Section 2.6]. Now the oxygen solubility remains effectively constant, but the partial pressure is greater than it is at the surface. To correct for this automatically involves practically the same procedure as has just been described for salting-out, except that the depth rather than the conductivity has to be monitored and the correction term which the function generator has to produce is the exponential in eq. (2.30) instead of that in eq. (5.27).

Other high-pressure effects may, however, cause more serious problems. The compressibility of water is $\sim 5 \times 10^{-5}$ atm^{-1} so a 1% change in solution volume will require a pressure of 10^3 atm. Such a pressure is found at depths in the region of 10 km and therefore it is unlikely that any significant change in electrolyte thickness will occur at great depths. Similarly, plastics commonly used for MPODs also have compressibilities in the region of 5×10^{-5} atm^{-1} (3, Ch. 9), so measurements should not be affected by external pressure on the membrane. But if there are air pockets present in the electrolyte then comparatively small pressure changes can cause large fluctuations in the electrolyte film thickness. The mounting and holding of the membrane in place then becomes an important aspect of the design of a detector (10). A simple solution to this problem is to use in addition to the gas permeable membrane a nonpermeable membrane over some other part of the electrolyte reservoir so that the internal and external pressures are equalized (63).

Another problem of measurements at great depth is the transmission of the signals to the surface free from loss and disturbance. One way of dealing with this is to submerge the amplifying electronics with the detector, but a more elegant solution is the use of telemetry (cf. 70, p. 22).

5.3.5.   POISONING

At the beginning of Chapter 4 the MPOD was introduced as an electrochemical sensor which could be used in polluted environments without detrimental effects on the electrode system. This is largely true for a wide

TABLE 5.5   Effect of H_2S on a MPOD with a Ag Cathode (55)

	Without metal salt film		With metal salt film	
	O_2	$O_2 + H_2S$	O_2	$O_2 + H_2S$
Current (μA)	230	205	250	245

range of applications for the membrane is a very effective barrier against dissolved and solid impurities. But whenever there are dissolved gaseous impurities these are not prevented from permeating the membrane and contaminating the system. The most disturbing effect of this type is the poisoning of the electrodes, particularly by sulfurous gases such as H_2S, SO_2, and thio-organic materials; Ag is especially sensitive. Table 5.5 gives data for a galvanometric detector with a silver cathode and lead anode in the absence and presence of H_2S (55), and the effect of the gas is readily seen.

Pt is also susceptible to poisoning by sulfur compounds (1,56), although less so than Ag. Au appears to be the least affected and most resistant (9), but even with Au poisoning eventually sets in and erratic results are obtained. In addition with Au there is still the problem of the attack of the anode, especially if this is Ag.

There are a number of ways of dealing with the problem of poisoning by sulfurous gases. An obvious solution is to remove them from the test solution before measurements are made. This can be achieved by, for example, oxidizing the impurities to less nocuous materials (e.g. H_2S to S), but one has to ensure that such an oxidation does not disturb the equilibrium oxygen concentration. Often, however, it is not expedient or convenient to pretreat the sample in this way, particularly for measurements *in situ*. A very simple but effective alternative to complete removal of the polluting gas from the test solution is to remove it from the gases diffusing through the membrane by having a "filter" between the inner membrane surface and the electrodes. Schmid and Mancy (55) have described a system which removes H_2S. Figure 5.25 illustrates the principle of its operation. A lens tissue soaked in a solution of a heavy metal salt—cadmium nitrate—is sandwiched between two membranes and removes H_2S by precipitating cadmium sulfide. The method is quite effective as illustrated by the comparative results in Table 5.5. But in very polluted conditions the metal salt solution will be rapidly used up and may even need replenishing between measurements, although prolonged life can be achieved by including an

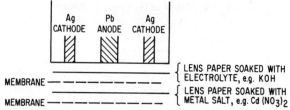

Fig. 5.25   Multilayer membrane system for H_2S removal. (Redrawn from reference 55.)

oxidizing agent, such as potassium persulfate, in with the metal salt. Another disadvantage of this system is that the multilayer membrane will considerably reduce the sensitivity and response of the detector, and so other methods are in general to be preferred.

Replacement of the metal cathode by a metal sulfide has been suggested (57). In particular, nickel sulfide is quite an effective catalyst for oxygen reduction (22) and clearly will not suffer from the problem of poisoning by sulfur compounds. A disadvantage of this solution, though, is that most metal sulfides are readily oxidized by air and this leads to a mixed oxide-sulfide electrode and problems of instability and residual currents arising from oxide reduction. However, a sulfide electrode can, with care, function satisfactorily, as illustrated in Fig. 5.26.

Another suggestion has been the use of a sulfide electrolyte (58). This ensures a constant reference potential by forming a sulfide electrode as the anode at the outset, and if one uses Au as a cathode then the resistance of the detector to polluted environments is high.

Finally, instead of taking preventive measures, remedial measures can be used. By superimposing an alternating current on the steady-state dc polarization, adsorbed impurities can be desorbed on the positive half cycle (59). To get useful results it is necessary to remove the ac current during the

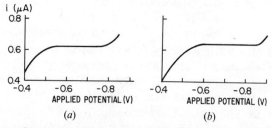

Fig. 5.26   Current-voltage curves for a clean nickel electrode and one exposed to H_2S (9). (a) Polished nickel electrode. (b) Nickel electrode after immersion in saturated hydrogen sulfide solution for four days.

measurement period as otherwise rectification currents are added to the oxygen reduction current. Instead of using an ac signal an anodic pulse can be applied periodically (60), and results for polluted systems are claimed to agree with the Winkler method.

None of the methods discussed here are really satisfactory solutions for *in situ* measurements to the problems of gaseous poisoning, for, although they all reduce the effects of contaminations, in the long run there is no alternative to replacing the electrolyte and cleaning the electrodes by mechanical polishing.

5.4. SOURCES OF ERROR

Although throughout this discussion of MPODs we have noted the advantages of such detectors, the need for caution and care in using them and in interpreting the results has also been emphasized. Figure 5.27 summarizes, for convenience, the possible sources of error that can occur and the manner in which they appear; reference to the appropriate section will suggest methods of treatment. Provided that attention is paid to these points then the measurement of dissolved oxygen with a membrane-covered detector can give accurate, precise, and reliable results very efficiently and economically.

5.5. APPLICATIONS

The published work on dissolved oxygen measurements using MPODs is voluminous, and rather than attempt to review this huge area we simply refer to a number of publications which give a comprehensive coverage of the field. Measurement of oxygen in respiratory air, blood, and tissues has been extensively covered in the reports of three symposia (70, 71, 74), while Fatt (73) has provided an extensive description of oxygen sensors for clinical applications. Hoare (2, Ch. VI) has reviewed more general applications of oxygen electrodes to biological systems and has included references to the use of MPODs for determination of oxygen in natural waters, sewage waters, blood, cells and tissues, photosynthesis studies, biological cultures, and other media. Spiehler (34) has added to and extended this coverage in a bibliography of more than six hundred papers, although the most recent papers in this bibliography date from pre-1970.

From these references and from what we have said throughout this chapter it is clear that MPODs have a number of distinctive and attractive features for dissolved oxygen measurements. They are suitable for use in a wide variety of environments, in particular for solutions containing ionic and

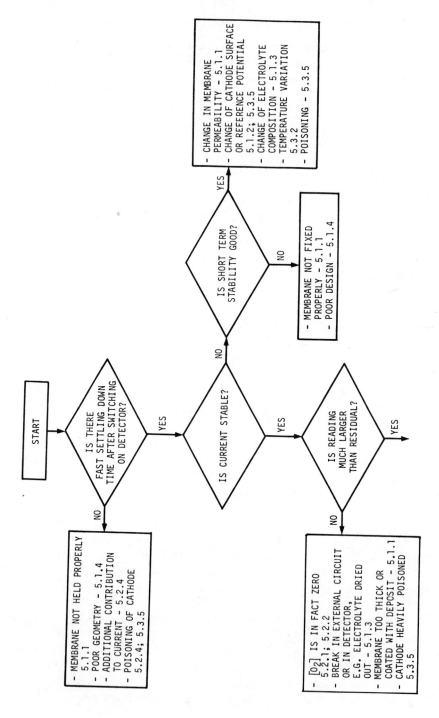

START

IS THERE FAST SETTLING DOWN TIME AFTER SWITCHING ON DETECTOR?

NO →
- MEMBRANE NOT HELD PROPERLY 5.1.1
- POOR GEOMETRY - 5.1.4
- ADDITIONAL CONTRIBUTION TO CURRENT - 5.2.4
- POISONING OF CATHODE 5.2.4; 5.3.5

YES ↓

IS CURRENT STABLE?

NO →

IS SHORT TERM STABILITY GOOD?

YES →
- CHANGE IN MEMBRANE PERMEABILITY - 5.1.1
- CHANGE OF CATHODE SURFACE OR REFERENCE POTENTIAL 5.1.2; 5.3.5
- CHANGE OF ELECTROLYTE COMPOSITION - 5.1.3
- TEMPERATURE VARIATION 5.3.2
- POISONING - 5.3.5

NO →
- MEMBRANE NOT FIXED PROPERLY - 5.1.1
- POOR DESIGN - 5.1.4

YES ↓

IS READING MUCH LARGER THAN RESIDUAL?

NO →
- [O$_2$] IS IN FACT ZERO 5.2.1; 5.2.2
- BREAK IN EXTERNAL CIRCUIT OR IN DETECTOR, E.G. ELECTROLYTE DRIED OUT - 5.1.3
- MEMBRANE TOO THICK OR COATED WITH DEPOSIT - 5.1.1
- CATHODE HEAVILY POISONED 5.3.5

YES ↓

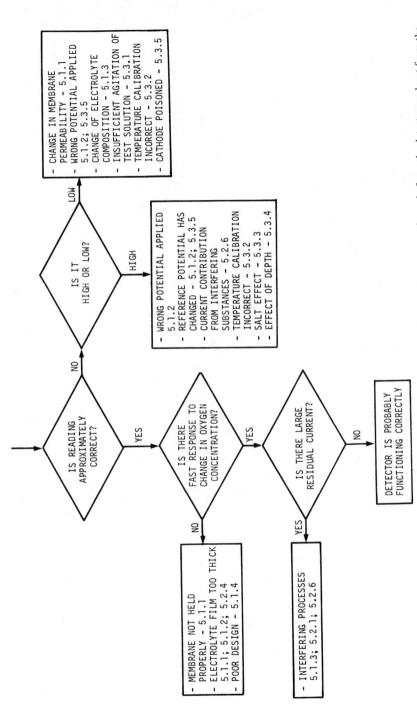

Fig. 5.27 Sources of error likely to be encountered with a MPOD. This chart assumes that errors arise only from the detector and not from the electronics associated with it.

119

organic contaminants (61), and they lend themselves readily to continuous monitoring *in situ*. Further, for environmental studies they monitor the activity of the dissolved oxygen, which is often a more meaningful measure than the concentration. There are, inevitably, difficulties associated with the use of membrane-covered electrodes, but very often they are associated with a lack of understanding of the procedure and principles of operation. This chapter has endeavored to throw some light on these problems.

REFERENCES

1. D. T. Sawyer, R. S. George, and R. C. Rhodes, "Polarography of gases," *Anal. Chem.*, **31**, 2 (1959).

2. J. P. Hoare, *The Electrochemistry of Oxygen*, Interscience, New York, 1969.

3. D. W. Van Krevelen, *Properties of Polymers*, Elsevier Publishing Company, Amsterdam, 1972.

4. H. Yasuda and W. Stone, "Permeability of polymer membranes to dissolved oxygen," *J. Polymer Sci.*, **4**, 1314 (1966).

5. E. C. Peirce, "Diffusion of oxygen and carbon dioxide through Teflon membranes," *A.M.A. Archives of Surgery*, **77**, 938 (1958).

6. C. Rogers et al., "Studies on the gas and vapour permeability of plastic films and coated papers," *Tappi*, **39**, 741 (1956).

7. K. H. Mancy, D. A. Okun, and C. N. Reilley, "A galvanic cell oxygen analyser," *J. Electroanal. Chem.*, **4**, 65 (1962).

8. J. R. Neville, "Oxygen sensor," U.S. Patent 3,071,530, issued Jan. 1, 1963.

9. M. L. Hitchman, unpublished results.

10. D. P. Lucero, "Design of membrane covered polarographic detectors," *Anal. Chem.*, **41**, 613 (1969).

11. I. Bergman, "Improvements in or relating to membrane electrodes and cells," U.S. Patent 1,200,595, issued July 29, 1970.

12. J. M. Hale, personal communication.

13. B. E. Conway, *Theory and Principles of Electrode Processes*, Ronald, New York, 1965.

14. W. M. Latimer, *The Oxidation States of the Elements and Their Potentials in Aqueous Solutions*, Prentice-Hall, New Jersey, 1961.

15. J. O'M. Bockris and A. K. N. Reddy, *Modern Electrochemistry*, Plenum Press, New York, 1970.

16. A. Damjanovic, in *Modern Aspects of Electrochemistry*, (J. O'M. Bockris and B. E. Conway, Eds.), No. 5, Butterworths, London, 1969.

17. J. Jacq and O. Bloch, "Electrochemical reduction of oxygen in hydrogen peroxide solution," *Electrochim. Acta*, **15**, 1945 (1970).

18. W. J. Albery and M. L. Hitchman, *Ring Disc Electrodes*, Clarendon Press, Oxford, 1972.

19. J. L. Newman, *Electrochemical Systems*, Prentice-Hall, New Jersey, 1973.

20. M. W. Breiter, *Electrochemical Processes in Fuel Cells*, Springer Verlag, Berlin, 1969.

21. D. T. Sawyer and E. T. Seo, "Reduction of dissolved oxygen at boron carbide electrodes," *J. Electroanal. Chem.*, **3**, 410 (1962).

22. A. K. M. S. Huq and A. J. Rosenberg, "Electrochemical behavior of nickel compounds," *J. Electrochem. Soc.*, **111**, 270 (1964).

23. V. T. Stack, "Dissolved oxygen probe," U.S. Patent 3,372,103, issued Mar. 5, 1968.

24. R. Briggs and M. Viney, "The design and performance of temperature compensated electrodes for oxygen measurement," *J. Sci. Instr.*, **41**, 78 (1964).

25. F. J. H. Mackereth, "An improved galvanic cell for determination of oxygen concentration in fluids," *J. Sci. Instr.*, **41**, 38 (1964).

26. E. I. Konnik and I. L. Mordukhovich, "Investigations on galvanic analysers for oxygen," *Soviet Electrochemistry*, **5**, 1233 (1969).

27. T. M. Doniguian, "Membrane type polarographic oxygen sensors," U.S. Patent 3,758,398, issued Sept. 11, 1973.

28. L. W. Niedrach and W. H. Stoddard, "Sensor with anion exchange resin electrolyte," U.S. Patent 3,703,457, issued Nov. 21, 1972.

29. F. Helfferich, *Ion Exchange*, McGraw-Hill, New York, 1962.

30. H. J. R. Maget, in *Handbook of Fuel Cell Technology*, (C. Berger, Ed.), p. 425. Prentice-Hall, New Jersey, 1968.

31. L. Niedrach and W. T. Grubb, in *Fuel Cells* (W. Mitchell, Ed.), Ch. 6, Academic, New York, 1963.

32. H. A. Liebhafsky and E. J. Cairns, *Fuel Cells and Fuel Batteries*, Ch. 14, Wiley, New York, 1968.

33. G. Halpert, A. C. Madsen, and R. T. Foley, "Analysis for oxygen by gas phase polarography at low temperatures and pressures," *Rev. Sci. Instr.*, **35**, 950 (1964).

34. V. Spiehler, "A bibliography of application areas for Beckman polarography oxygen analysers," Beckman Technical Report No. 545.

35. E. W. Molley, "Polarographic sensor," U.S. Patent 3,406,109, issued Oct. 15, 1968.

36. R. W. Pittman, "A diffusion controlled cathode for measurement of oxygen tensions," *Nature*, **195**, 449 (1962).

37. K. H. Mancy, *Instrumental Analysis for Water Pollution Control*, Ann Arbor Science Publishers Inc., Ann Arbor, Mich., 1971.

38. C. J. Koch and J. Kruuv, "Measurement of very low oxygen tensions in unstirred liquids," *Anal. Chem.*, **44**, 1258 (1972).

39. H. Lipner, L. R. Witherspoon, and V. C. Champeaux, "Adaptation of a galvanic cell for micro-analysis of oxygen," *Anal. Chem.*, **36**, 204 (1964).

40. R. G. Clem, "Rotated platinum cell for controlled potential coulometry," *Anal. Chem.*, **43**, 1853 (1971).

41. J. F. Reynolds, "Comparison studies of Winkler *vs* oxygen sensor," *J. Water Poll. Contr. Fed.*, **41**, 2002 (1969).

42. *Standard Methods for the Examination of Water and Wastewater*, American Public Health Association, 13th ed., 1971.

43. J. Crank, *The Mathematics of Diffusion*, Oxford University Press, Oxford, 1956.

44. D. E. Carritt and J. W. Kanwisher, "An electrode system for measuring dissolved oxygen," *Anal. Chem.*, **31**, 5 (1959).

45. V. T. Stack, "Agitator for dissolved oxygen probe," U.S. Patent 3,360,451, issued Dec. 26, 1967.

46. J. Hissel and J. Pire, "Microdetermination of dissolved oxygen in boiler feed water by coulometry at constant current," *Bull. Centre Belge et Document Eaux*, **44**, 76 (1959).

47. J. Crank and G. S. Park, *Diffusion in Polymers*, Academic, New York, 1968.

48. D. W. van Krevelen, *Properties of Polymers*, Elsevier Publishing Company, Amsterdam, 1972.

49. E. Keonjian and J. S. Schaffner, "Shaping of the characteristics of temperature sensitive elements," *Trans. Amer. Inst. Elec. Eng.*, **73**, 396 (1954).

50. C. R. Droms, "Thermistors for temperature measurements," in *Temperature— Its Measurement and Control in Science and Industry*, Vol. 3, Part 2 (A. I. Dahl, Ed.), Reinhold Publishing Corporation, New York, 1962.

51. H. V. Larson et al., "Method of linearising thermistor thermometer data in calorimetry," *J. Sci. Instr.*, **38**, 400 (1961).

52. *Application Manual for Computing Amplifiers*, George A. Philbrick Researches Inc., Boston, 1966.

53. J. I. Smith, *Modern Operational Circuit Design*, Wiley-Interscience, New York, 1971.

54. J. G. Graeme, G. E. Tobey, and L. P. Huelsman, *Operational Amplifiers—Design and Applications*, McGraw-Hill, New York, 1971.

55. M. Schmid and K. H. Mancy, "The electrochemical determination of dissolved oxygen in water in the presence of hydrogen sulphide," *Chimia*, **23**, 398 (1969).

56. T. Loucka, "Adsorption and oxidation of sulphur and sulphur dioxide at the platinum electrode," *J. Electroanal. Chem.*, **31**, 319 (1971).

57. M. L. Hitchman, W. Mehl, and J. P. Millot, "Electrochemical oxygen detector," U.S. Patent 3,785,948, issued Jan. 15, 1974.

58. E. J. Amdur, "Electrochemical sensor," U.S. Patent 3,515,658, issued June 2, 1970.

59. R. A. Olson, F. S. Brackett, and R. G. Crickard, "Oxygen tension measurement by a method of time selection using the static platinum electrode with alternating potential," *J. Gen. Physiol.*, **32**, 681 (1949).

60. L. V. Kamlyuk and A. K. Kamlyuk, "Determination of oxygen dissolved in water by a polarographic method with the aid of a solid electrode with electrochemical depolarisation," *Dokl. Akad. Nauk. Beloruss. SSR.*, **11**, 368 (1967).

61. C. Legler, "Contribution to the determination of dissolved oxygen in water," *Limnologica*, **4**, 291 (1966).

62. C. Boutin et al., "Comparative study of an electrochemical method and the chemical method of Winkler for determination of oxygen dissolved in sea water," *Cah. Oceanogr.*, **21**, 555 (1969).

63. A. E. Wheeler, "In situ oxygen and carbon dioxide sensors for oceanography," *J. Aircraft*, **2**, 436 (1965).

64. C. E. W. Hahn, A. H. Davis, and W. J. Albery, "Electrochemical improvement of the performance of pO_2 electrodes," *Respiration Physiology*, **25**, 109 (1975).

65. F. A. Keidel, "Coulometric analyser for trace quantities of oxygen," *Indust. Eng. Chem.*, **52**, 490 (1960).

66. W. J. Wingo and G. M. Emerson, "Calibration of oxygen polarographs by catalase catalysed decomposition of hydrogen peroxide," *Anal. Chem.* **47**, 351 (1975).

67. W. L. Masterton, "Salting coefficients for gases in seawater from scaled particle theory," *J. Soln. Chem.*, **4**, 523 (1975).

68. J. D. Borkowski and M. J. Johnson, "Long-lived steam sterilisable membrane probes for dissolved oxygen measurement," *Biotech. and Bioeng.*, **9**, 635 (1968).

69. H. Enoch and V. Falkenburg, "An improved membrane system for oxygen probes," *Soil Sci. Soc. Am. Proc.*, **32**, 445 (1968).

70. F. Kreuzer, *Oxygen Pressure Recording in Gases, Fluids, and Tissues*, S. Karger, Basel, 1969.

71. M. Kessler et al. (Eds.), *Oxygen Supply*, Urban and Schwarzenberg, Munich, 1973.

72. J. Izydorczyk, W. Misniakiewicz, and K. Raszka, "Working conditions of a Clark-type sensor with a polyethylene membrane in the presence of vapours of chlorine derivatives of methane," *J. Electroanal. Chem.*, **70**, 365 (1976).

73. I. Fatt, *Polarographic Oxygen Sensors*, CRC Press, Cleveland, 1976.

74. F. Kreuzer and H. P. Kimmich, "Recent developments in oxygen polarography as applied to physiology," in *Measurement of Oxygen* (H. Degn, I. Balslev, and R. Brook, Eds.), Elsevier Scientific Publishing Company, Amsterdam, 1976.

75. B. B. Lloyd and B. Seaton, "Storage effects in an oxygen electrode," *J. Physiol.*, **207**, 29P (1970).

6

MEMBRANE-COVERED POLAROGRAPHIC DETECTORS—NONSTEADY-STATE MEASUREMENTS

"Measure time, control it! Do everything on time! Save time, make time count, work fast!"

From a leaflet distributed by the Soviet Time League

6.1. INTRODUCTION

In the chapter dealing with the theory of the current for a MPOD three equations were presented corresponding to very short times [$< \sim 0.1$ s— eq. (4.12)], short times [$< \sim 10$ s—eq. (4.13)], and long times [eq. (4.14)] after the detector is switched on. Subsequently we obtained the expression for the steady-state current [eq. (4.17)] and then largely restricted the discussion to this case. However, although steady-state signals are particularly easy to handle and analyze, there are a number of real advantages to be gained by making nonsteady-state measurements, especially if the time of measurement is restricted to the period either before the diffusion layer at the electrode has reached the membrane surface or before it has moved a large part of the way through the membrane.

Let us consider again eq. (4.12). As was done previously this equation is reduced to a simplified form for $t < 0.1$ s

$$i_{L,t} = nF \frac{K_b}{K_0} C_s \left(\frac{D_e}{\pi t} \right)^{1/2} \tag{4.15}$$

We also have noted that this equation is of the same form as eq. (3.38) for the time-dependent, limiting current at a stationary electrode when diffusional transport only occurs. One of the advantages of measuring this current instead of the steady-state, limiting current is that the high current at short times gives increased sensitivity. But for MPODs further advantages accrue from the fact that at these short times the current is independent of the membrane. First, any changes in the properties of the membrane and any deposits on the membrane will have no effect on the measurement. Second, the temperature coefficient of the detector is now no longer dependent on the permeability coefficient of oxygen in the membrane but on its diffusion coefficient in the electrolyte, and this is a much better behaved

124

parameter; the temperature correction, for example, can be done automatically with more confidence. Third, since the diffusion layer has not spread out to the outer surface of the membrane while a measurement is taken there is no influence of solution flow on the reading obtained; stirring of the test solution is not necessary. Finally, a measurement can be made very rapidly and so the overall life of the detector should be increased. The membrane and problems associated with it have occupied much of the discussion so far, but now by simply making measurements at short times we have effectively removed all concern about the membrane, other than the fact that it must still be there to protect the electrodes against gross pollution.

However, there are two additional complexities with nonsteady-state measurements at such short times. One is the increased complexity in making measurements at times of less than one second. This problem can be reduced by taking readings at slightly longer time intervals, say of the order of several seconds, when chart recorders can be used and current values can be read off at leisure. At these longer times the diffusion layer will have spread into the membrane, but by no means to the outer edge of the membrane [cf. Fig. 4.5]. This means that the transport will now be controlled by the membrane properties, as indicated by eq. (4.16):

$$i_{L,t} = nFC_s \left(\frac{P_m}{\pi t} \right)^{1/2} \tag{4.16}$$

Thus some of the advantages of very short time measurements have been lost, although since the diffusion layer is well within the membrane the effects of stirring will still be minimal. But even with measurements made in times of the order of seconds, for completely automated monitoring it is clear that electronic sampling techniques will also be required. Thus, on balance, it would seem preferable to make measurements at times of less than one second and gain the advantages that go with such measurements, even though it does involve a little more circuitry.

The other complexity with nonsteady-state measurements arises from the fact that the current transients are in general rather slower than expected on the basis of eqs. (4.15) and (4.16) [Section 5.2.4]. Nevertheless, as will be seen in the next section, it is still possible to use transient currents in an empirical fashion (1).

6.2. TECHNIQUES AND RESULTS

Figure 6.1 illustrates the principle of the nonsteady-state measurement. The MPOD is taken from open circuit to the application of a potential across the cell, either voltammetrically or galvanometrically. After the first few milliseconds, when current is being used to charge the double layer at

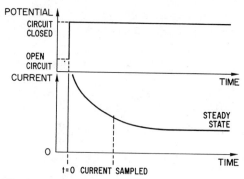

Fig. 6.1 Principle of nonsteady-state measurements.

the electrode—electrolyte interface, the measuring procedure can be started. The most straightforward method of doing this is to record the total current versus time curve and then to analyze it in the manner described earlier [Section 3.6.2].

Figure 6.2 shows some experimental *i*—*t* curves. The time scale is in seconds and so the current will be determined by transport in the membrane. In this instance, therefore, the current will show membrane-dependent char-

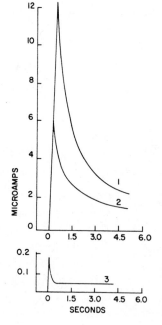

Fig. 6.2 Experimental current transients for a MPOD (1). Curve 1—11.5 ppm O_2 at 25°C; curve 2—8.2 ppm O_2 at 25°C; curve 3—residual current. (Reproduced by permission of Elsevier Sequoia S.A.)

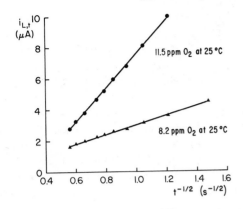

Fig. 6.3   Plot of data from Fig. 6.2.

acteristics, but one would still expect, on the basis of eq. (4.16), a linear dependence between the current and the inverse of the square root of time. Figure 6.3 plots this relationship for data taken from Fig. 6.2 and good straight lines are obtained.

Readings obtained at fixed time intervals for different oxygen concentrations allow a calibration plot to be made—Fig. 6.4. Putting the steady-state calibration curve on the same plot shows very clearly the increase in sensitivity obtained from nonsteady-state measurements.

After the time interval when a measurement has been made has passed, the cell can be returned to open circuit in readiness for the next measurement. Lilley et al. (1) found that in their case 99% reequilibration had been

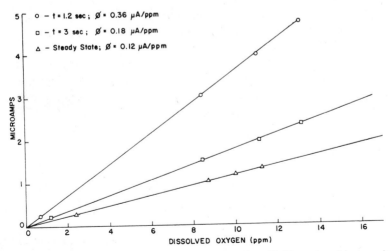

Fig. 6.4   Calibration plot for nonsteady-state measurements (1) (Reproduced by permission of Elsevier Sequoia S.A.)

achieved within 5 min. This is much longer than one would expect on the basis of theory, and the reasons for this sluggish behavior have been discussed earlier [Section 5.2.4]. The slow return to equilibrium, however, need not prevent measurements being made at more frequent intervals, for provided that the time interval between pulses is maintained constant reproducible results are obtained (1).

The effects of stirring were also studied by Lilley. The stirrer was operated at full speed with the propellor blades about 2 cm below the detector membrane. The first detectable deviation between the curves obtained in quiescent and stirred solutions occurred at 4.5 s after the voltage was applied. At this point the deviation was 3%. After 7.5 s the deviation was 8.3% and after 12 s it was 25%. These results are much as one would expect from the calculation of the time required for the diffusion layer to spread toward the outer edge of the membrane. With measurements made before about 4 s have passed the output of the detector is independent of any convection in the system being analyzed, and so calibration curves obtained in a quiescent system can be applied to rapidly flowing systems as well.

Instead of recording a complete current—time curve and analyzing it afterward, a more convenient procedure is to monitor the current at a predetermined time and then to hold this value to be read at leisure. Readings obtained in this way can be used directly to make, or to compare with, a calibration plot. Another method of analysis is to sample the current at a number of chosen times and to store the information prior to analysis. This essentially combines the complete record and single point methods, and while being very convenient it requires a small computer or a microprocessor to handle the data this way.

A simpler method of making use of more than just one current value for each measurement is to measure the charge between two instances in time. The principle of this technique is illustrated in Fig. 6.5 and the advantages are twofold. One advantage is that the electronic circuitry is somewhat simpler than that required for a sample and hold. The second advantage is that in measuring charge any noise in the signal can be "averaged" out quite readily.

Others (2–5; 6, p. 99) have also reported work on nonsteady-state measurements with a MPOD. Mancy and Schmid (2, 3) have obtained similar results to those of Lilley and co-workers (1), while Fowler and Oldham (4) have described transient experiments with a sensor in which the cathode is some distance behind the membrane, thus ensuring membrane-independent characteristics. The Fowler and Oldham system in fact corresponds simply to the case of transport to a stationary electrode by diffusion alone [Section 3.6.2], and because of the well-behaved properties of such transport in aqueous solution the sensitivity is reproducible from one cell to another

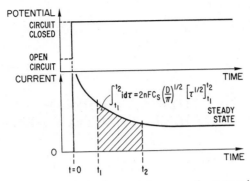

Fig. 6.5 Principle of analysis of current-time curves by current integration.

and also agrees very well with theory; thus the necessity for calibration and recalibration is avoided. However, this new sensor has the serious drawback of being extremely slow in responding to oxygen changes, so although it is an interesting device it would seem to have no real advantage over pulsed measurements with a normal sensor and readings being taken before the diffusion layer has spread into the membrane.

Apart from the work quoted here, very little further work on nonsteady-state measurements with MPODs seems to have been done. Because of the advantages we have mentioned it would be profitable for further studies to be made of the technique.

REFERENCES

1. M. D. Lilley, J. B. Story, and R. W. Raible, "The chronoamperometric determination of dissolved oxygen using membrane electrodes," *J. Electroanal. Chem.*, **23**, 425 (1969).
2. M. Schmid and K. H. Mancy, "The electrochemical determination of dissolved oxygen in water with the help of a modified membrane electrode technique," *Schweiz. Zeit. Hydrol.*, **32**, 328 (1970).
3. K. H. Mancy, "In situ measurement of dissolved oxygen by pulse and steady-state voltammetric membrane electrode systems," *Electrochem. Soc. Extended Abstr.*, Vol. 75–1, Abstr. No. 328, 1975.
4. J. K. Fowler and K. B. Oldham, "Voltammetric membrane cell used in the equilibrium mode for dissolved oxygen assay," *Electrochem. Soc. Extended Abstr.*, Vol. 75–1, Abstr. No. 329, 1975.
5. K. Kunze, "Continuous absolute oxygen pressure measurements with short impulses," *Pflugers Arch. ges. Physiol.*, **283**, 4 (1965).
6. M. Kessler et al., Eds., *Oxygen Supply*, Urban and Schwarzenburg, Munich, 1973.

OTHER METHODS OF MEASUREMENT—ELECTROCHEMICAL

"Like—but oh how different!"

W. Wordsworth, *Yes, it was the mountain Echo!*

7.1. INTRODUCTION

Having looked at membrane-covered polarographic detectors in some detail we now consider other methods that are currently used for dissolved oxygen measurements. A number of general reviews and comparison studies have been made of methods for determining oxygen concentrations in water (1–8, 69, 70), and, in addition, regular annual (9) and biennial (10) literature surveys appear covering the latest developments in the field. In this chapter we shall deal with electrochemical techniques and in the following chapter all of the other common methods. In neither case, though, do we attempt to give a detailed coverage, but rather a general review and comparison; reference should be made to the original articles and papers listed in the reviews just quoted for more information.

7.2. POTENTIOMETRIC METHODS

The principle of potentiometric measurement (11, Ch. 3) is based on the Nernst equation [Section 3.2]:

$$E_e = E^0 + \frac{RT}{nF} \ln\left(\frac{C_\infty^O}{C_\infty^R}\right) \tag{3.9}$$

In the case of oxygen reduction in aqueous alkaline solution C_∞^O corresponds to molecular oxygen, O_2, and C_∞^O to either HO_2^- or OH^- depending on whether a two- or four-electron reaction is involved. However, as has been seen [Section 5.1.2], not only is it difficult to observe the theoretical, reversible potential values for the oxygen electrode reaction, but also the values that are found are very dependent on electrode and solution pretreatment and history. Thus monitoring of dissolved oxygen in aqueous media by direct potentiometry is not very practicable, although not impossible (78, 79).

130

Measurements can be made indirectly by potentiometry though, if the oxygen is first removed from solution into the gaseous phase and then monitored using a rather special electrochemical system. This system can be represented by the cell

$$p'_{O_2}, Pt \quad \left| \quad \begin{matrix} \text{solid oxide ion conducting} \\ \text{electrolyte} \end{matrix} \quad \right| \quad Pt, p''_{O_2}$$

where p'_{O_2} and p''_{O_2} represent the oxygen partial pressures at the cathode and anode, respectively. The electrolyte is a solid which has an appreciable conductivity for the oxygen ion. There are no known materials which have such a conductivity at ambient temperatures, but at elevated temperatures (600–1600°C) a number of suitable materials exist (12, 13). One of these materials is zirconia, ZrO_2, or more usually stabilized zirconia (for example, $ZrO_2.x$ CaO) and this has predominantly oxygen ion conduction over a range of oxygen pressures from above 1 atm to about 10^{-25} atm (depending on temperature). The electrodes are porous and are at equilibrium with different oxygen partial pressures as indicated. At open circuit the voltage of the cell indicates the ratio of oxygen chemical potentials at the electrodes according to the equation

$$E = \frac{RT}{4F} \ln\left(\frac{p''_{O_2}}{p'_{O_2}}\right) \tag{7.1}$$

provided that the electrolyte exhibits exclusive ionic conduction between the two electrodes over the measured oxygen pressure range. And, unlike the O_2/OH^- couple, the O_2/O^{2-} couple at elevated temperatures is reversible and is not subject to the problem of solution impurities or electrode history; at the high temperatures used a uniform oxide film is formed on the platinum electrodes. If p''_{O_2} is considered to be a time-independent and known reference electrode (e.g. air with $p''_{O_2} = 0.21$ atm) then the cell can indicate the unknown or variable p_{O_2} with high accuracy.

Figure 7.1 illustrates schematically a zirconia electrolyte system (14). Degassing the sample is done by one of the methods described in [Section 8.2.1], and the gaseous oxygen is dried and pretreated before passing it into the cell. A pair of short (1 cm) porous Pt electrodes (III and IV) are placed downstream and form the open circuit cell for the continuous monitoring of the oxygen activity of the gas. A pair of longer (22 cm) porous Pt electrodes (I and II) serve as an upstream "pumping" cell. This pumping cell can be used either to generate known concentrations of oxygen coulometrically in order to calibrate the system, or to pump residual oxygen out of the inert carrier gas; for the latter purpose it is possible under favorable conditions to reduce the residual oxygen level to the order of 10^{-20} atm (14). The constriction mixer is present to ensure good gas mixing just prior to monitoring, and the thermocouple to measure the monitoring temperature.

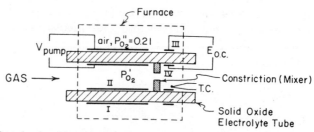

Fig. 7.1 Zirconia electrolyte system (14). V_{pump}—applied potential for "pumping" out the system; $E_{o.c.}$—open circuit potential measurement; T.C.—thermocouple. (Reproduced by permission of the publisher, The Electrochemical Society Inc.)

There are a number of advantages to the high-temperature, potentiometric measurement. First, the cell is essentially a solid-state device, and second, because of the logarithmic response, a wide range of oxygen partial pressures are readily accessible. On the other hand, the need to remove oxygen from solution before analysis could be inconvenient. Also the oxygen may need to be cleaned after degassing from solution since at the cell operating temperature a chemical equilibrium between oxygen and any oxidizable gases present (for example, NO, NH_3, SO_2, and CO) may be established, and then the oxygen partial pressure measured will be that after oxidizable species have reacted; in other words, the cell potential will be dependent on the partial pressures of the oxidizable gases and on the equilibrium constants for their reaction with oxygen. The high temperature required for measurement is clearly a disadvantage, even with a well-insulated and controlled furnace. In addition, while the method could be the basis of on-site monitoring, it cannot be used *in situ*. Finally, the fact that there is a logarithmic relationship between the parameter actually measured, voltage, and the oxygen partial pressure, while, as mentioned above, allowing a wide range of oxygen partial pressures to be accessible, nevertheless means that small changes in p_{O_2} will give even smaller variations in output, and stable and accurate electronic amplification will be necessary—a change of one order of magnitude in p_{O_2} produces only a change of about 50 mV in the output of the cell. The properties and physical principles of zirconia-based oxygen sensors have been reviewed recently (83–85), and the application of these cells to various situations has been described (19, Ch. VI; 56; 83; 85); commercial instruments are available.

7.3. AMPEROMETRIC METHODS

The measurement of a current which is related in a relatively simple and reproducible way to the concentrations of the electroactive species—in this case oxygen—forms the basis of amperometric monitoring [Sections 3.4 and 3.5]. The MPOD makes use of this principle, but here we are concerned with systems where the electrodes are not separated from the test solution by a gas permeable membrane. Some common types of electrode system that can be used for amperometric analysis were briefly described in Section 3.5, and these systems can be divided into two main classes—the dropping mercury electrode (DME) and solid electrodes.

7.3.1. DROPPING MERCURY ELECTRODE

A typical arrangement for this electrode system is given in Fig. 7.2. A capillary about 5–10 cm long and of about 0.05 mm bore diameter is positioned vertically and mercury is allowed to flow through it under a pressure of 30–60 cm at a rate of 1–5 mg s^{-1}, so that the lifetime of a drop lies between 2 and 5 s. With the very simple potential divider circuit a varying voltage can be applied to the cell; the reference electrode in this case is also the auxiliary electrode separated from the main cell by a sintered disk and agar plug. With the arrangement of Fig. 7.2 the current is measured with a galvanometer, but generally more sophisticated electronics are used.

In the presence of a reducible species a current—voltage curve, or polarogram, is obtained—Fig. 7.3. The characteristic sigmoid shape is what one would expect on the basis of the earlier discussion on i—V curves [Section

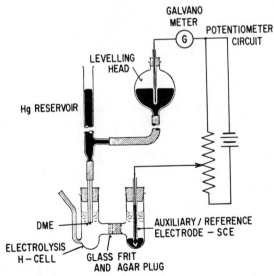

Fig. 7.2 A diagram of a dropping mercury electrode.

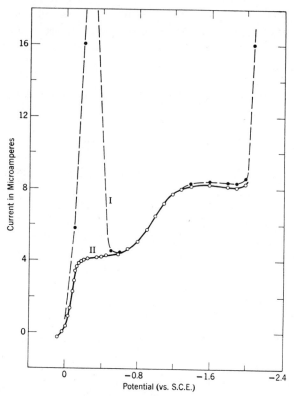

Fig. 7.3 Current-voltage curve for oxygen reduction at a DME. Curve I—in the absence of a maximum suppressor: curve II—in the presence of 0.01% thymol. (Reproduced by permission from reference 15. Copyright by the American Chemical Society.)

3.4], and a double wave is found for oxygen reduction corresponding to reduction initially to H_2O_2 and then to H_2O at higher potentials [Section 5.1.2]. Measurement of the limiting current on either plateau gives a measure of the dissolved oxygen concentration, although generally the second wave is used since the current is larger and a current maximum, if present (see below), does not interfere with the measurement.

The current at a dropping mercury electrode, unlike that found with a MPOD, is not a genuine steady-state current since the concentration of the electroactive species is varying both with distance and time as the mercury drop electrode grows in size and eventually falls off. When transport controlled conditions prevail (that is, the electrode potential is sufficiently large for the surface concentration of the reacting species to be zero [Section 3.4]) then we can write for the current [cf. eq. (3.30)]

$$i_L = nFAD\frac{C_\infty}{\delta}$$ (7.2)

where A is the electrode area and δ is again the diffusion layer thickness. The area of the drop, assuming it to be spherical, is simply $4\pi r^2$, and r, the radius, can be readily obtained from knowing the steady flow rate of mercury through the capillary. Letting this flow rate be m g s^{-1}, then the volume of a given drop at any time, t, after the initial formation of the drop is

$$\frac{4\pi r^3}{3} = \frac{mt}{\rho}$$

where ρ is the density of mercury in g cm^{-3}. Hence r is obtained. The diffusion layer thickness, δ, is given by the same equation as that used for diffusion to a plane electrode [eq. (3.40)] since for a thin layer the curvature of the drop can be ignored. Substituting for A and δ in eq. (7.2) gives

$$i_L = 4\pi nF\left(\frac{3mt}{4\pi\rho}\right)^{2/3} D\frac{C_\infty}{(\pi Dt)^{1/2}}$$

$$= knF\left(\frac{m}{\rho}\right)^{2/3} D^{1/2}t^{1/6}C_\infty$$

where k contains all the numerical constants. This equation is the famous Ilkovic equation. It predicts that the current rises with the power of $t^{1/6}$. However, the current will not continue to rise indefinitely as after a few seconds the drop falls off, the current falls sharply and then starts to increase again as a new drop grows. Thus the characteristic saw-tooth pattern of a polarogram is obtained. The polarogram in Fig. 7.3 does not show this characteristic pattern since it is based on single point, averaged measurements.

In addition to the normal sigmoid current—voltage curve the polarogram in Fig. 7.3 shows a large current maximum, beginning at 0 V (vs. SCE) and returning to the limiting current just before the second wave starts. Such maxima are often found on polarograms at the DME and the types of maxima that occur and the theories presented to explain their existence are discussed in most polarography texts (e.g. 17, Ch. X; 18, Ch. 3). Since maxima can interfere with the measurement of diffusion currents it is obviously desirable to prevent their occurrence, and frequently this is possible by addition to the solution of maximum suppressors. These are substances which, at quite low concentration, allow normal curves to be obtained, and they include substances such as sugar, indicators, dyestuffs, and gums which are normally nonreducible. Some suppressors are specific to a particular reducible species or to a limited potential range, while others appear to be effective over most potentials and with most electroactive species. Curve II of Fig. 7.3 was taken in the presence of a trace of thymol and in this case the two waves are clearly visible since the maximum is completely suppressed without affecting the shape of the second wave significantly. Other suppressors may remove the maximum but not leave the second wave unaffected (21).

One of the great advantages of the DME is that the electrode surface is renewed every few seconds, and such an arrangement minimizes any long-term effects of adsorption of impurities and also allows correction for the short term effects. To be offset against this advantage are the problems associated with adapting the technique and making it robust enough to be used for field work. Furthermore, in many instances, particularly in biological studies, the DME cannot be used for *in situ* measurements because of the danger of mercury dissolving and contaminating or poisoning the system. Despite these very real problems the DME has been used fairly extensively for dissolved oxygen analysis. Inevitably, there have been various modifications to the basic method. One such modification that has been quite successful is the wide bore, rapid DME (71, 72, 80). This design, as the name suggests, has a capillary with a large internal diameter—~ 0.8 mm— and as a result the flow rate is considerably enhanced. The advantages arising from a rapid rate of mercury flow are that the deleterious effects of pollutants on the electrode are minimized even further, and that the current sensitivity is significantly increased. Hoare (19, Ch. VI) has made a useful survey of applications of the DME in the analysis of natural waters, sewage wastes, and biological systems. Several points are worth noting.

Even though we have emphasized the fact that the renewal of the electrode surface every few seconds minimizes the effects of impurities on the electrode reaction, nevertheless with *in situ* measurements in heavily polluted waters considerable care still has to be taken to ensure that the current is not enhanced or otherwise disturbed by interfering substances (20, 73, 74). An interfering substance need not be electroactive, for readily adsorbed surface active agents can affect both the magnitude of the diffusion current and the position of the waves (21). The conductivity of the sample may also be too low for polarographic analysis [cf. Section 3.3.1], in which case a sufficient quantity of a suitable salt must be added. It is interesting to note, though, that one problem not encountered in *in situ* measurements, especially those made in sewage wastes, is that of current maxima since naturally occurring suppressors are often present (20).

Polarography at a DME is a highly developed technique and in addition to the classical dc method which we have outlined, there are a whole range of ac methods. Also measurements can be made differentially by taking the derivative of the current or by recording the current just before the end of the drop life—Tast polarography (17, 18). Very little use seems to have been made of these more advanced techniques for oxygen analysis, although all of these are aimed at reducing the effects of impurities and/or increasing the sensitivity. The sensitivity limit of classical polarography is $10^{-5}-10^{-6}M$ (that is, 0.05–0.1 mg liter^{-1}) and the precision is about ± 0.02 mg liter^{-1}.

Derivative polarography, Tast polarography, square wave polarography, and ac polarography can all extend the lower limit to $10^{-6}-10^{-7}M$, while pulse polarography allows analysis down to $10^{-7}-10^{-8}M$. One application of derivative polarography (22) has shown it to be twice as sensitive as the Winkler method, but there appears to be considerable scope for using some of the other nonclassical techniques.

7.3.2. SOLID ELECTRODES

The main disadvantages of the DME that have been mentioned are the problems associated with making it robust enough for field work and the danger of contaminating the system under study by the mercury dissolving. Solid electrodes allow the construction of rigid, robust monitoring equipment and do not usually have the problem of dissolution and sample contamination. One type of solid electrode that has been considerably used in dissolved oxygen analysis is the rotating platinum electrode. This system typically consists of a platinum wire protruding several millimeters through the wall of a glass tube and the whole assembly being rotated at a constant speed, usually in the range 5–50 Hz. The limiting current is, as usual, directly related to the dissolved oxygen concentration, but, as mentioned earlier, the exact form of the equation for the current is not readily calculable from first principles; the rotating wire electrode has to be used in a semiempirical fashion.

Eden and co-workers (82) have described the production of large numbers of miniature, stationary solid electrodes by means of integrated circuit techniques, rather in the same manner suggested for preparing multicathode Clark-type sensors (86, p. 33) [Section 5.1.4]. However, unlike MPODs of course, such solid electrodes will achieve no true steady state [Section 3.5] and so measurements would have to be made at short times. If one wishes to use a solid electrode for dissolved oxygen monitoring then probably a rotating disc electrode (23) would be preferable because of the well-defined and calculable transport that this electrode imposes on the test solution. Again, as with other types of solid electrode, adaptation to *in situ* oxygen measurements would be relatively easy. Adams (23) has described in general the types of solid electrode listed in Table 3.2.

The circuitry necessary to drive a solid electrode voltammetric device could be of the very simple kind shown in Fig. 7.2 for a DME, or of the more extensive sort shown in Fig. 5.20 for a MPOD; in the latter case the temperature compensation circuit would have to be different because of the absence of the membrane. Of course, just as with MPODs, it is possible to have galvanic systems as well as voltammetric ones. A number of such galvanic cells have been described (1, 70, 75, 76). Hoare (19, Ch. VI) has reviewed the application of solid galvanic and voltammetric devices to dissolved oxygen analysis in general, and Fatt (86) to the analysis of biological systems.

Although monitoring equipment based on solid electrodes is inherently easier to adapt to *in situ* measurements than that using a DME, some of the problems of the DME still remain. So, for example, the sample must have sufficient conductivity to minimize effects of migration, and also the current can be significantly changed from its true value by interfering substances. Solid electrodes, however, suffer from an additional drawback in that the electrode surface, once contaminated, is not so readily renewable as the DME. In fact, this problem of surface poisoning is so serious with solid electrodes that they can be used only in fairly clean environments. Thus, Ingols (24) reported that stationary and rotating platinum electrodes cannot be used for detection of oxygen in sewage and sludges because the electrode surfaces become filmed with impurities, rendering them inoperative. It has been suggested that a square wave superimposed on the steady dc potential (25) or a reverse current pulse of variable amplitude and length (26) be used to remove the adsorbed impurities by oxidation, but it is doubtful whether these cleaning procedures will have a marked effect on thick deposits of contaminants. Perhaps a combination of such periodic, anodic electrode treatments together with a continuous scraping of the surfaces of both anode and cathode with an insulating scraper (27, 57, 58) might provide a complete solution to the problem of electrode poisoning. However, there then seems to be some delay in obtaining stable readings from the system.

In conclusion, we can say that the only really satisfactory way to use amperometric methods for oxygen analysis is to remove the oxygen from the sample and then redissolve it in a clean electrolyte solution where the analysis can be made directly (19, Ch. VI; 28) or indirectly (29). Under these conditions the sensitivity will be comparable to that of classical polarography and with care calibration need only be done infrequently; electrolytic generation of oxygen provides a convenient means of calibration (30, 31).

7.4. COULOMETRIC METHODS

Coulometry is based on eq. (3.1)

$$w = \frac{Mit}{nF} \tag{3.1}$$

where w is the mass (in grams) of substance of atomic or molecular weight M that is transformed electrochemically by the current i (in amps) passing for a time t (in seconds). Since there is a direct proportionality between w and it the determination of the quantity of electricity, or charge, may be used as an "absolute method" of chemical analysis, depending not on standard samples but on the absolutely defined quantities, the ampere and the second. However, to use eq. (3.1) in this way it is necessary that the current efficiency at the working electrode be 100%—that is, no side reactions should occur at the electrode—and that the coulombic efficiency, η—the fraction of the total amount of electroactive substance in the sample that undergoes reaction at the working electrode—should also be 100%. So, for example, in the Hersch cell (32–34), which we describe later in this section, the current efficiency is generally 100% for oxygen reduction, but the coulombic efficiency falls rapidly with increasing flow of the sample [Fig. 7.4] and only at flow rates less than 2 ml min^{-1} does eq. (3.1) apply. At higher flow rates a coulombic yield less than 100% necessitates a calibration with known samples, although Keidel (30) has described a similar cell to that of Hersch, but in which complete conversion of the oxygen is achieved at flow rates of 100 ml min^{-1} and higher. But even though, in general, the Hersch cell has a coulombic yield considerably less than unity at high flow rates, it is still very much greater than that obtained with a polarographic detector. Thus at 100 ml min^{-1}, η is $\sim 10\%$ for the Hersch cell compared with $\sim 0.1\%$ for a typical measurement made with a MPOD with $i \sim 10\ \mu A\ cm^{-2}$ and operating in a sample of 100 ml of air-saturated water for a period of 10 min. In both cases the current efficiency is 100%, but in the Hersch cell the percentage of oxygen in the sample that is reduced is much greater than that reduced with the MPOD, and this, as we shall see, arises essentially because of the design of the Hersch system which allows much more efficient transport of the oxygen to the electrode surface.

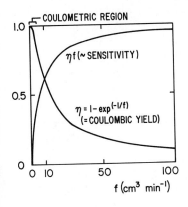

Fig. 7.4   Effect of flow rate on the coulombic yield of a Hersch cell. (Reproduced by permission from reference 33. Copyright by the American Chemical Society.)

There are two basic ways of performing a coulometric analysis. One is by passing a constant current through the cell (35) and the other is by applying a controlled potential to one electrode in the cell (36). The second method is clearly very similar to the use of the MPOD and other amperometric methods that have been described here, but the main difference arises, as we have just noted, in the coulombic yield that is obtained. Polarography can quite legitimately be considered as a coulometric process, but it is more convenient to treat it from the point of view of amperometry, if only because the very low efficiency means it would be difficult to calibrate the system at such yields. The term *coulometry*, therefore, while originally being reserved for those cases where eq. (3.1) holds, now generally includes also cases where the coulombic yield during the course of the measurement is greater than ~1%. Below this level the depletion of the test sample can be regarded as negligible and the height of a diffusion controlled current plateau can be used for analysis; a current measurement is made. Above this level depletion is significant, the current falls steadily, and analysis must be done by current integration; a charge measurement is made. More information on the two basic methods of coulometry, as well as their application to analysis, can be found in various places (e.g. 11, Ch. VI; 37, Chs. 12 and 14; 38; 39).

For oxygen analysis several coulometric devices have been suggested, all based on controlled potential coulometry. Hersch was the first to describe a coulometric cell (32). This cell consisted of a Pt-wire spiral cathode placed in a 24% KOH solution and with a lead amalgam anode. Dissolved oxygen could be estimated with the cell by degassing the solution and bubbling the carrier gas and sparged oxygen through the cell over the Pt cathode. The system served as an efficient oxygen detector, and increased sensitivity was observed if the Pt cathode was only partially submerged instead of being completely covered by the alkaline electrolyte.

Later, Hersch (33) designed a cell in which the Pt cathode was replaced by Ag. In this cell—Fig. 7.5—a Pb foil anode is wrapped around a steel tube and the PVC film is saturated with a solution of about $5M$ KOH. The gaseous test sample enters the cell, travels down the annular space between the cathode and the glass housing, where the oxygen is reduced, and leaves the cell via the central channel. The electrochemical reactions occurring at the two electrodes are basically the same as those found in a galvanic cell with a Ag cathode and Pb anode [Section 5.1.2], but, of course, the coulombic yield is much greater here.

The coulombic yield of the Hersch detector can be readily related to the flow rate of the gas sample. If we consider the gas entering the cathode compartment, the current will take a value proportional to the initial concentration, C_{in}, of the oxygen; let this current be $i_{L,in}$. As the gas stream passes through the cathode compartment the oxygen concentration falls and

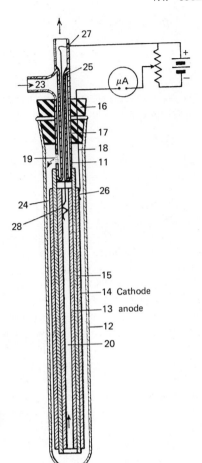

Fig. 7.5 The Hersch cell (34). Key: 11—double walled glass tube; 12—cell housing; 13—anode; 14—cathode; 15—PVC diaphragm separating electrodes; 16—stopper to seal cell; 17—Pt contact to anode; 18—steel contact to cathode; 19—port for gas; 20—central channel; 23—gas entry; 24—annular space; 25—gas exit; 26—contact to cathode; 27—Pt/glass seal; 28—contact to anode. (Reproduced with permission.)

as a result so does the current. For controlled potential coulometry at an electrode in an electrolyte solution the fall of the current is given by (37, Ch. 12)

$$i_{L,t} = i_{L,\text{in}} \exp(-\beta' t)$$

where $i_{L,t}$ is the current after time t, and β' equals $DA/V\delta$; V is the volume of solution, A the electrode area, D the diffusion coefficient, and δ the diffusion layer thickness. For the Hersch cell V/t can be replaced by the gas flow rate, f cm^3 s^{-1},

$$i_{L,t} = i_{L,\text{in}} \exp\left(-\frac{\beta A}{f}\right) \tag{7.3}$$

with β representing D/δ. The coulombic yield, η, is given by

$$\eta = \frac{C_{in} - C_{out}}{C_{in}} = 1 - \frac{C_{out}}{C_{in}} \tag{7.4}$$

where C_{in} and C_{out} are, respectively, the oxygen concentrations at the inlet and outlet of the cell. Because of the proportionality between current and concentration eq. (7.4) can be written as

$$\eta = 1 - \exp\left(-\frac{\beta A}{f}\right) \tag{7.5}$$

Figure 7.4 gives a plot of this equation for βA equal to unity. This is not an unreasonable value since with $D \sim 10^{-5}$ cm^2 s^{-1} and $\delta \sim 10^{-3}$ cm, A would then be 10^2 cm^2; and in the cell of Fig. 7.5 it is ~ 93 cm^2. Equation (7.5) and Fig. 7.4 both show that coulombic yield falls rapidly with increased flow rate. On the other hand, this effect can be counteracted by having a large surface area electrode. In fact, at any given flow rate one can, in principle, bring the yield up to unity simply by increasing the electrode area.

The large surface area of a coulometric cell is essentially what distinguishes it from a polarographic device, for it is this feature which helps to give the cell its high coulombic yield. But it is not only a question of increasing the surface area. Hersch (33) observed that the current for a given gas flow at a partially submerged cathode was greater than if the electrode was totally immersed. He explained this phenomenon in terms of oxygen being adsorbed on a dry electrode surface and, in the adsorbed state, diffusing toward the visible meniscus before being cathodically reduced.

More detailed studies by Will (40) of this effect showed, however, that this mechanism contributes insignificantly to the total current and that most of the electrode reaction takes place in a thin (10^{-5}–10^{-3} cm) invisible film of electrolyte above the meniscus. A similar explanation has been found to be true (19, Ch. VIII) for other electrodes exposed to gases, such as porous electrodes. A large electrode surface area is thus important, but also the ratio of electrode surface to electrolyte volume must be large for high coulombic yields. Keidel (30) achieves a large surface to volume ratio by bubbling the gas through a porous cathode, but a porous membrane electrode such as is used in fuel cells can serve just as well (81).

In addition to increasing the coulombic yield of a cell, making the electrode surface area large improves the sensitivity of the cell. The sensitivity, s (in μA/ppm by volume in the gas), is derived from Faraday's law and is given by

$$s = 4.462 \times 10^{-5} nFf\eta \tag{7.6}$$

As f, the flow rate, increases, s also increases if η remains constant, which can only occur if A increases also. At a given flow rate the maximum sensitivity is clearly obtained with $\eta = 1$, and again in order to achieve this the electrode area must be increased. But if nothing is done to compensate the fall in yield with increased flow then a saturation value of sensitivity is obtained. Substituting eq. (7.5) into (7.6) we have

$$s = 4.462 \times 10^{-5} nFf \left[1 - \exp\left(-\frac{\beta A}{f} \right) \right] \qquad (7.7)$$

For $\beta A/f \ll 1$, or $f \gg \beta A$, the exponential term can be expanded:

$$s \simeq 4.462 \times 10^{-5} nFf \left[1 - \left(1 - \frac{\beta A}{f} \right) \right]$$

$$= 4.462 \times 10^{-5} nFA\beta \qquad (7.8)$$

and s is seen to be independent of f; this is shown in Fig. 7.4.

With a coulombic yield of unity the sensitivity of a Hersch cell is seen, from eq. (7.6), to be $17.2 \ \mu A \ (ppm)^{-1} (cm^3 \ s^{-1})^{-1}$ at N.T.P.; this is for oxygen concentrations in ppm by volume in the gaseous phase. Since the residual current in the absence of oxygen is about $1 \ \mu A$ this means the lower limit of measurement is ~ 0.1 ppm. If the coulombic yield is down then this limit will not be so low. For measurements of small concentrations it is therefore necessary to try to keep η high, as has already been noted.

At the other end of the scale, the upper limit of measurement of Hersch cells is generally well below $1\% \ O_2$ (by volume), and the limitation arises from the response curve, as a function of oxygen concentration, flattening out. The range of linearity can however be extended by bifurcating the gas stream so that one minor branch leads directly into the cell while the major part passes through an oxygen scavenging unit and then into the cell (33). This limitation at the upper end is important for dissolved oxygen measurements, since a concentration of, say, 8 mg liter^{-1} corresponds to $\sim 5 \ cm^3$ of gaseous oxygen, which if sparged by, say, one liter of inert gas constitutes a 0.5% concentration in the gas phase. Thus some calculation is necessary before a coulometric cell is used in this way for dissolved oxygen analysis.

With a coulombic yield of unity no calibration of a Hersch-type device is in principle necessary, although it is worthwhile to check that the efficiency is still 100% from time to time. When the yield is less than unity a calibration curve must be made using gases of known oxygen content. These may be conveniently prepared using electrolytic generation (30, 31, 33); the curve is linear from below 1 ppm to 0.01% and beyond. With a system where either $\eta = 1$ or where a calibration curve has been made the precision of coulometric detectors is good and is limited mainly by accuracy of sample flow and current measurement; an overall error of ± 0.1 ppm is achieved without too much difficulty (30).

The speed of response of a Hersch cell to sudden changes of oxygen level is of the order of 30 s for the total change (33), but this does not include time lags in the sample feed lines. If these lags are included then typically 2–3 min are required for 90% response, and 4–5 min for 95% response (30, 41). Kendall (41) has pointed out that poor response characteristics are primarily due to the presence of large volumes in the components preceding the cell and in the cell itself, and to the presence of a relatively large volume of electrolyte which provides a high resistance to oxygen mass transfer. These problems are overcome by reducing the amount of dead volume in the system and by having the cathode, in the form of silver coated glass beads, separated from the anode by a sheet of filter paper soaked in electrolyte and extending into an electrolyte reservoir—Fig. 7.6. Using such a system the time for complete response to a sample is reduced to 15 s, as Fig. 7.7 shows.

It is worth noting that in Fig. 7.7 the curves were obtained with a discrete sample of gas, and so a classic coulombic analysis is made with the peak area (the charge) being proportional to the oxygen content of the gas. Very often though a continuous sample is used, and then a current is taken as

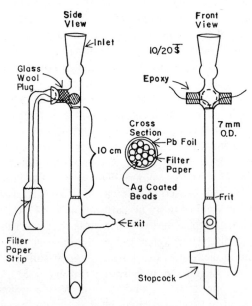

Fig. 7.6 Modified Hersch cell for fast response. (Reproduced by permission from reference 41. Copyright by the American Chemical Society.)

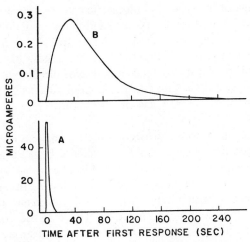

Fig. 7.7 Comparison of fast response and standard Hersch cells. Curve A—silver coated beads as cathode in Hersch cell. Curve B—Delphi B2503-8 Hersch cell. Carrier gas flow rate = 100 ml min^{-1}; change in oxygen concentration = 0 to 100 ppm. (Reproduced by permission from reference 41. Copyright by the American Chemical Society.)

the measure of oxygen present; this is because a balance is kept between the rate at which electrochemical transformation takes place and the rate of gaseous influx. This latter type of analysis is commonly called continuous coulometry, but the expression *derivative coulometry* has been suggested as an alternative (31) since this shows that the method is coulometric and yet that a current (the time derivative of charge) is measured.

One great advantage of coulometric devices is that temperature changes have no effect on the current obtained per oxygen molecule passing through the cell [cf. eq. (3.1)]. However, the volume of a given sample mass is directly proportional to the absolute temperature and the sensitivity calculated above has to be corrected

$$s_T = 17.2\left(\frac{273}{T}\right) \mu A \text{ (ppm)}^{-1} \text{ (cm}^3 \text{ s}^{-1})^{-1} \tag{7.9}$$

Alternatively, the sensitivity can be maintained constant and the sample flow rate adjusted by about 0.37% per °C at temperatures above 273°K. But even though some form of temperature compensation is needed, it is still considerably less than that required for a MPOD [Section 5.3.2] and this means it is simpler to achieve. The temperature variation is also likely to be more reproducible for a coulometric cell than for a MPOD since the former device does not have problems of temperature hysteresis in a membrane.

A disadvantage of the type of coulometric cells that we have described is that they cannot be used for *in situ* measurements of dissolved oxygen. So the oxygen has to be stripped from the water sample [Section 8.2.1] and then passed into the analyzer cell. Figure 7.8 gives a flow diagram for liquid analysis using a Keidel cell (30). Since an alkaline electrolyte is preferred for coulometric analysis [cf. Section 5.1.3] acid gases such as CO_2 should either be scrubbed out of the carrier gas or else a bicarbonate electrolyte used (30, 33).

There is no reason, however, why coulometric analysis for oxygen should not be performed directly on a liquid sample, and Eckfeldt and Shaffer (31) have designed a cell for this purpose—Fig. 7.9. The liquid sample enters the bottom of the cell and the cathode compartment, having first passed through an electrolyte injector which makes the test solution basic and of sufficient conductivity for the coulometric reaction. The cathode consists of small diameter (0.7–0.8 mm) silver metal spheres with a total surface area in excess of 600 cm². The anode is a Cd wire in the form of an open-wound helix and with a geometric area of about 50 cm²; this auxiliary electrode is immersed in its own electrolyte compartment. With flow rates from less than 1 cm³ min⁻¹ to more than 6 cm³ min⁻¹, and with solutions of oxygen concentration in the range 0.03–40 ppm, coulombic yields close or equal to 100% are obtained. The precision of the experiments is generally ~1% or better and

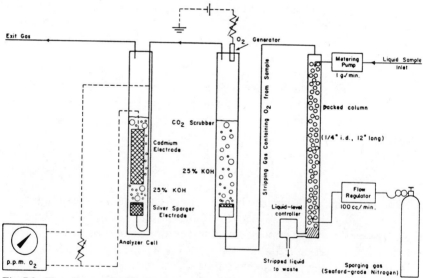

Fig. 7.8 Flow diagram for dissolved oxygen analysis with a coulometric cell. (Reproduced by permission from reference 30. Copyright by the American Chemical Society.)

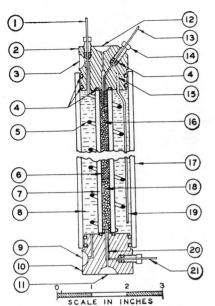

1. Lead wire to secondary electrode
2. Upper Kel-F end structure
3. Nitrogen gas exit port
4. Buna N rubber O-rings
5. Secondary electrode, cadmium $^1/_{16}$ inch dia., 30 inches long
6. Glass wool packing
7. Porous ceramic diaphragm tube
8. Secondary electrolyte
9. Nitrogen gas inlet port
10. Lower Kel-F end structure
11. Sample solution inlet port
12. Sample solution outlet port
13. Lead wire to second section of working electrode
14. O-ring sealing nut
15. Secondary electrolyte inlet (plan, item 9)
16. Second section of working electrode
17. Cell casing, borosilicate glass pipe, $1\frac{1}{2}$-inch I.D.
18. First section of working electrode
19. Kel-F inlet tube for secondary electrolyte
20. Silver tubing conductor to first section of working electrode
21. Leadwire to silver tubing conductor

Fig. 7.9 Coulometric dissolved oxygen analysis cell. (Reproduced by permission from reference 31. Copyright by the American Chemical Society.)

good agreement with other independent measurements is found. The speed of response of the cell is as good as that reported by Kendall (41) for the improved Hersch cell system, and the temperature sensitivity is less than that for a Hersch cell simply because of the much lower volume expansivity of the liquids. The real disadvantage of a system such as this for direct coulometric analysis of dissolved oxygen is that, like solid electrodes in amperometric analysis, electrode contamination and poisoning quickly occur unless precautions are taken to clean up the solution sample beforehand. A MPOD with a large cathode area and operating in a small sample volume could function efficiently as a coulometric cell and would overcome this problem, yet still retain the advantage of insensitivity to temperature changes.

Hersch cells have been largely used for oxygen analysis in gaseous samples [e.g. in petrochemical plant stream (42)], but some use has been made of them for dissolved oxygen analysis, for example, in boiler waters (33). If efficient and reproducible stripping of oxygen from a sample can be maintained then they are interesting candidates for on-site monitoring with a very long life (33), requiring little attention apart from replenishing of the electrolyte (33, 41). A number of commercial systems are available.

7.5. CONDUCTOMETRIC METHODS

Conductometric methods of analysis for dissolved oxygen can be conveniently divided into two main types. Those methods which are based on the formation of ionic species by oxygen leading to an increase in the conductivity of the test solution, and those methods which depend on the change of resistance of a metal oxide with oxygen partial pressure. In the first type of analysis classical conductometry in solution [cf. Appendix M] is carried out and two reactions have been generally used—the formation of nitrite ions from nitric oxide and oxygen, and the formation of metal ions by reaction between a metal and oxygen. The second type of analysis is less conventional in that resistance measurements are made on solid metal oxides; cobalt oxide (CoO) has been mainly used in this technique.

7.5.1. WITH ION FORMATION IN SOLUTION

The reaction involving the addition of nitric oxide gas to the oxygenated water sample (44) is as follows:

$$4NO + O_2 + 2H_2O \rightarrow 4H^+ + 4NO_2^-$$ (7.10)

The formation of nitrous acid results in an increase in electrical conductance which is, from eqs. (M.4) and (7.10), directly proportional to the dissolved oxygen concentration. A flow diagram for analysis by this method is given in Fig. 7.10. Ions already present in the sample are removed by passage through a deionizing column and the background or residual con-

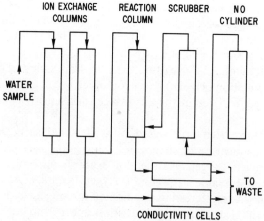

Fig. 7.10 Flow diagram for analysis by conductivity measurement in solution.

ductivity is monitored. The deionized water enters the reaction column and the conductivity is measured with a second conductivity cell. The change in conductivity (or the difference if the sample is bifurcated prior to the reaction occurring and the conductivities of the reacted and unreacted samples are measured in parallel) is readily expressed directly as a dissolved oxygen concentration.

The formation of metal ions in solution is illustrated by the reactions between lead (45) or thallium (46–52) and oxygen:

$$2Pb + O_2 + 2H_2O \rightarrow 2Pb^{2+} + 4OH^- \tag{7.11}$$

$$4Tl + O_2 + 2H_2O \rightarrow 4Tl^+ + 4OH^- \tag{7.12}$$

In the case of Tl oxidation, which is the most commonly used, the conductivity of a 10 μM solution of TlOH is about 2.7×10^{-6} ohm^{-1} cm^{-1}. Therefore the change observed in the solution resistance in a cell of length 1 cm and cross sectional area 1 cm^2 when 80 μg liter^{-1} of O$_2$ is present in the test sample will be 370 kΩ, which is readily detectable. The lower limit of detection is obviously determined by the residual conductivity of the sample and the resolution of the equipment. Ordinary deionized water in equilibrium with carbon dioxide in the air has a conductivity $\sim 0.7 \times 10^{-6}$ ohm^{-1} cm^{-1} and this alone would fix a lower limit of measurement ~ 20 μg liter^{-1}. However, by making a differential measurement an order of magnitude improvement can be achieved and the lower limit extended to ~ 2 μg liter^{-1}; several reports have appeared describing measurements at these low levels (47, 51, 52). The upper limit of measurement is about 1000 μg liter^{-1}, or 1 mg liter^{-1}, of dissolved oxygen. This is mainly determined by the fact that at this concentration a solution ~ 0.1 mM in TlOH is produced and above this level the linearity between specific conductance and concentration begins to break down. In any case, once 1 mg liter^{-1} is reached there are many other more convenient methods available.

The precision of conductometric methods is of the order of a few percent, which is really quite good when one considers the low level of measurement. Time response to step changes in dissolved oxygen concentrations is typically about a few minutes for 95% of the final reading. The temperature sensitivity [Appendix M] gives rise to a change in reading $\sim 2\%/°C$, which is significant enough to necessitate careful temperature control or compensation.

Overall, conductivity methods involving ion formation in solution can be seen to be capable of very high sensitivity and to be suitable for continuous monitoring. They are not, however, suitable for *in situ* measurements, partly because of the problem of possible electrode contamination by impurities; although if the technique of high-frequency conductometry (11, Ch. 2), with either external electrodes to the cell or no electrodes at all, could be extended down to the concentration levels that we have discussed, this problem would no longer be present. But in addition to electrode contamination conductometric analysis has the drawback of needing sample pretreatment by ion exchange columns, which can cause loss of dissolved oxygen by adsorption on the large reaction surfaces and by activation of reactions between oxygen and dissolved organic and inorganic substances.

The major application of conductometric analysis is in determination of dissolved oxygen in boiler feed waters, boiler condensates, and other applications where high-purity, low-conductivity waters are used. Several commercial instruments that use the conductivity principle are available. Reviews of instruments and of the general area of dissolved oxygen analysis by conductometric methods have been published (48, 50).

7.5.2. WITH SOLID METAL OXIDES

Many metal oxides, and cobalt oxide in particular, show a dependence of conductivity on temperature and oxygen partial pressure. Pure cobalt oxide, CoO, is a semiconductor with very high resistivity. Deviations from stoichiometry, however, result in structural defects—cobalt vacancies—that strongly influence the electrical conductivity of the oxide. These properties make the conductivity of both single crystal and polycrystalline materials at high temperatures ($> 800°C$) depend on the ambient oxygen partial pressure (59–62). The use of this characteristic of CoO has been suggested as a means of monitoring oxygen (63, 64).

McIlwrick and Phillips (63) used the method of measuring the resistance of a CoO filament, first suggested by Duquesnoy and Marion (65, 66), to monitor the oxygen concentration in purified nitrogen streams. They calibrated their system at $1040°C$ with suitable mixtures of carbon monoxide and carbon dioxide which equilibrate very rapidly at this temperature ($2CO + O_2 \rightleftharpoons 2CO_2$) to give partial pressures in the required range. The relation between the logarithm of resistance and the logarithm of partial oxygen pressure (in atmospheres) is approximately linear, falling off at oxygen pressures less than about 10^{-8} atm. Using this method these workers were able to show that with manganese (II) oxide it is possible to reduce the oxygen concentration in nitrogen streams to better than 1 part in $10^{9.5}$.

Logothetis and co-workers (64) used CoO ceramics, prepared by standard ceramic techniques, in the determination of air-to-fuel ratios of an internal combustion engine. The dependence on the partial pressure of oxygen of the resistivity of a CoO ceramic in the temperature range $800–1000°C$ is shown in Fig. 7.11. Also included in the figure are data of Eror and Wagner (60) for single crystal CoO at $\geq 1000°C$; good agreement between the two

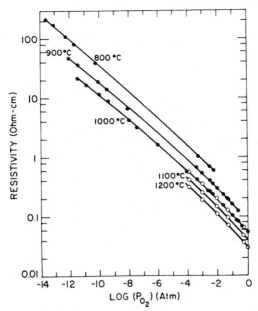

Fig. 7.11 Dependence of the resistivity of CoO on the oxygen partial pressure (64). Full circles are data points for results on ceramics; open circles are results for single crystals (60). (Reproduced by permission of the American Institute of Physics.)

types of materials is obtained. In the range between 800 and 1000°C the resistivity of the ceramic is proportional to $p_{O_2}^{-1/4}$ for high partial oxygen pressures, and it tends toward $p_{O_2}^{-1/6}$ for lower partial pressures. These results are in general agreement with measurements on single crystals for $\theta \geqq 1000°C$ (59–61).

Figure 7.12 shows one sensor configuration with a CoO ceramic. It essentially consists of the ceramic bar, with two platinum wire electrodes embedded in the ceramic, inside a small furnace which maintains the temperature of the sensor at 1000°C. The gas to be analyzed enters the sensor through a port in the alumina cover, diffuses through the porous CoO ceramic, and then leaves via another port. Using this type of sensor measurements were made on the exhaust gases of an internal combustion engine and partial pressures in the range of $1–7 \times 10^{-2}$ atm were found.

The work of Logothetis and co-workers shows, as does that of McIlwrick and Phillips, that measurement of the resistivity of CoO provides a versatile and sensitive means of monitoring oxygen. In particular it is found (64) that the behavior of the ceramics is relatively insensitive to chemical composition and gross structural details. So, for example, results do not depend markedly

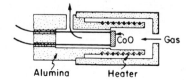

Fig. 7.12 Typical CoO sensor configuration (64). (Reproduced by permission of the American Institute of Physics.)

on the purity of the starting material, on the details of preparation of the ceramic, or on the handling of the material. These characteristics are, of course, very important for sensor applications. The temperature coefficient of resistance for CoO is small, about 0.3% per °C, so that a 5% accuracy in the determination of oxygen partial pressure requires a temperature control of about 4°C; this is not too difficult to achieve. Response time is also good with ceramics showing less than 1 s for step changes in oxygen concentrations.

Like conductometric analysis with ion formation in solution, the use of solid metal oxides provides a very sensitive method for continuous monitoring of oxygen. However, with solid metal oxides clearly the oxygen has to be stripped from the water before analysis, which has advantages and disadvantages, as noted in the discussion on potentiometric sensors [Section 7.1]. It is interesting in passing to note the difference between a potentiometric device, using zirconia as an electrolyte, and the sensors described here. Both obviously make use of solid metal oxides, but the former method relies upon ionic conduction in the oxide for the cell to function electrochemically, while the latter requires electronic conduction so that a pure ohmic dependence can be observed (67).

An important limitation on the usefulness of metal oxide sensors results from the fact that oxide decomposition can occur at the high temperatures used, either at very low oxygen partial pressures or in the presence of reducing gases; although for dissolved oxygen analysis it is possible that neither of these conditions would actually be present. To our knowledge no studies have been made on the use of metal oxide resistivity for dissolved oxygen measurements, but the technique is promising enough to warrant further investigation because of its high sensitivity. Other oxides should also be studied—for example, ceria (67) and zinc oxide (68). Zinc oxide is interesting since it shows an oxygen-dependent resistivity at ambient temperatures. In the absence of light it shows a very slow response to a step change in oxygen concentration (several hours), but when illuminated with an iodine-tungsten lamp 95% response is obtained in about 8 min; modifications in the design of the device may improve on this. The sensitivity of zinc oxide does not appear to be as good as that of CoO—the lower limit reported is 10^{-2} atm—and the response is also very sensitive to moisture variations (77), but a membrane-covered ZnO film at a constant humidity with a small light source might prove to be an interesting oxygen monitor.

7.6. ANOTHER METHOD

Recently a series of papers have appeared on the method of ionic transport in lanthanum fluoride thin films (53–55). At ambient temperatures these films—with thicknesses in the range $0.2-1.2$ μm—are good F^- anion conductors, having a dc resistivity $\sim 10^{12}$ ohm-cm when kept in 1 atm of nitrogen gas. If a cell is constructed by vapor deposition with LaF_3 as a solid electrolyte—Fig. 7.13—and is incorporated into an electrical circuit so that a fixed bias potential is applied, then upon exposure to a reducible gas the resistivity of a virgin cell decreases by two orders of magnitude or more. During a "break-in" period in the reducible gas the initial sensitivity decays in an exponential fashion, reaching a constant rate after several hours. Once a cell has been broken in the cell current becomes a relatively stable function of applied potential and reducible gas concentration in the ambient atmosphere.

Figure 7.14 shows current versus applied potential plots for a Bi (anode)/ (LaF_3)/Au (cathode) cell in 1 atm of N_2 and in 1 atm O_2. As the applied potential is increased a particular potential, V_c, is noted below which the cells are rather insensitive to oxygen, but above which sensitivity increases strongly with voltage. V_c is used as a characteristic potential and is defined as the difference between the applied potential giving zero current in oxygen and that where the oxygen current begins to increase rapidly. Defined in this way V_c is largely independent of anode material, but is found to correlate with the electron affinity of the gas being reduced. The important thing is, however, from a practical viewpoint, that above V_c the cell current varies as the sum

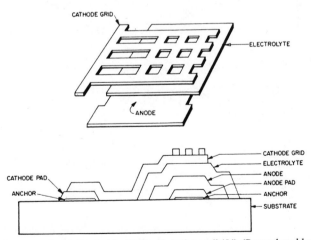

Fig. 7.13   Schematic of construction for LaF_3 electrolyte cell (54). (Reproduced by permission of the publishers, The Electrochemical Society Inc.)

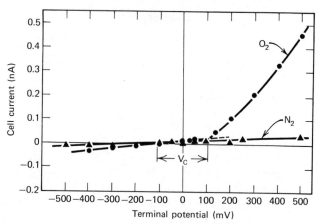

Fig. 7.14 Static current-potential relationship for a $Bi/LaF_3/Au$ cell in 1 atm of O_2 and 1 atm of N_2 (54). (Reproduced by permission of the publishers, The Electrochemical Society Inc.)

of linear and logarithmic terms in oxygen concentration. Above 20% of O_2 in mixtures of O_2 and N_2 the variation is linear with concentration and below this the logarithmic term causes significant curvature. Figure 7.15 shows current sensitivity versus % O_2 for cells with various anodes. Cells with Bi anodes are the most sensitive, but show the greatest low-concentration nonlinearity, while cells with Au anodes respond more linearly.

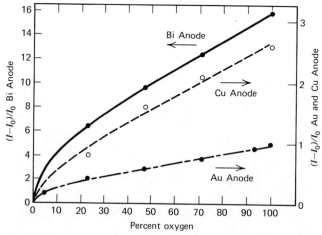

Fig. 7.15 Current sensitivity to oxygen of LaF_3 cells with Bi, Cu, and Au anodes with terminal potentials of 0.0, 0.4, and 0.4 V, respectively (54). (Reproduced by permission of the publishers, The Electrochemical Society Inc.)

The behavior of these cells is not clearly understood, but they show characteristics of both an amperometric and a conductometric device: amperometric since a definite electroreduction occurs at the cathode and a current flows; conductometric since the flow of current is accompanied by a nondiffusion limited transport through the electrolyte which gives rise to an ohmic dependence of current on the applied potential. The conduction mechanism is postulated to involve some form of oxygen anion, for example, O^- or O^{2-}. Whatever the mechanism, though, the cells are potentially interesting for dissolved oxygen analysis, especially since they are compact, sturdy, and will be cheap to make. The disadvantages foreseen are the interference from other reducible gases, the limited lifetime found at present for the cells, and the problem of breaking in each cell until there is only a further slight decrease in sensitivity.

REFERENCES

1. K. H. Mancy and T. Jaffe, "Analysis of dissolved oxygen in natural and waste waters," U.S. Public Health Serv. Publ., 999-WP-37, 1966.

2. V. P. Rothstein and V. N. Shemyakin, "Determination of oxygen in natural waters," Dokl. Otd. Kom. Geogr. Obs. SSSR., 2, 168 (1967).

3. H. Thiele, "Possibility of continuous measurement of oxygen content in water and sewage as well as oxygen consumption in systems with high consumption rate," Industrieabwasser, 1968, 22.

4. A. Borrowman, "Measurement of dissolved oxygen," Birmingham Univ. Chem. Eng., 21, 21 (1970).

5. K. H. Mancy, Ed., Instrumental Analysis for Water Pollution Control, Ann Arbor Science Publishers Inc., Ann Arbor, Mich., 1971.

6. H. Guerin and F. Girard, "Methods of determination of oxygen in gases," Analusis, 2, 562 (1973).

7. J. P. Masson, "Determination of dissolved oxygen in water," Analusis, 2, 608 (1973).

8. H. Noesel, "Modern technology for electrochemically measuring dissolved oxygen in aqueous media," Messtechnik, 81, 15 (1973).

9. "Annual Literature Reviews," Journal of the Water Pollution Control Federation.

10. "Reviews of Analytical Applications—Water Analysis," Anal. Chem., 1977, 1975, 1973, 1971, etc.

11. D. R. Browning, Ed., Electrometric Methods, McGraw-Hill, London, 1969.

12. J. Kummer and M. E. Milberg, "Ionic conduction and diffusion in solids," Chem. Eng. News, 1969, 90.

13. L. Heyne, "Some aspects of solid electrolytes," Electrochim. Acta, 15, 1251 (1970).

14. Y. K. Agrawal, D. W. Short, R. Gruenke, and R. A. Rapp, "The control of oxygen activities in argon-oxygen mixtures by coulometric titration," J. Electrochem. Soc., 121, 354 (1974).

15. I. M. Kolthoff and R. S. Miller, "The reduction of oxygen at the dropping mercury electrode," *J. Am. Chem. Soc.*, **63**, 1013 (1941).

16. I. S. Longmuir and F. Allen, "The polarographic determination of oxygen," *J. Polarog. Soc.*, **8**, 63 (1961).

17. I. M. Kolthoff and J. J. Lingane, *Polarography*, Interscience, New York, 1952.

18. D. R. Crow and J. V. Westwood, *Polarography*, Methuen and Company, London, 1968.

19. J. P. Hoare, *The Electrochemistry of Oxygen*, Interscience, New York, 1968.

20. E. W. Moore, J. C. Morris, and D. A. Okun, "The polarographic determination of dissolved oxygen in water and sewage," *Sewage Works J.*, **20**, 1041 (1948).

21. K. H. Mancy and D. A. Okun, "Automatic recording of dissolved oxygen in aqueous systems containing surface active agents," *Anal. Chem.*, **32**, 108 (1960).

22. S. T. Talreja, B. J. Bhalala, and P. S. Rao, "Determination of dissolved oxygen in sea water concentrates by derivative polarographic method," *Salt Res. Ind.*, **6**, 82 (1969).

23. R. N. Adams, *Electrochemistry at Solid Electrodes*, Marcel Dekker, New York, 1969.

24. R. S. Ingols, "Experience with solid platinum electrodes in determination of dissolved oxygen—Georgia Institute of Technology Studies," *Sewage Ind. Wastes*, **27**, 7 (1955).

25. R. A. Olson, F. S. Brackett, and R. G. Crickard, "Oxygen tension measurement by a method of time selection using the static platinum electrode with alternating potential," *J. Gen. Physiol.*, **32**, 681 (1949).

26. L. V. Kamlyuk and A. K. Kamlyuk, "Determination of oxygen dissolved in water by a polarographic method with the aid of a solid electrode with electro-chemical depolarisation," *Dokl. Akad. Nauk. Beloruss. SSR*, **11**, 368 (1967).

27. L. Kalman, "Apparatus for the electrochemical determination of oxygen," Swiss Patent 469,981, issued Apr. 30, 1969.

28. R. E. Meyer, F. A. Posey, and P. M. Lantz, "Electrochemical method for monitoring the oxygen content of aqueous streams at the ppb level," *Desalination*, **11**, 329 (1972).

29. T. Takahashi, H. Sakurai, and T. Sakamoto, "Continuous polarographic deter-mination of oxygen in water or gas by use of a zinc amalgam reductor," *Bunseki Kagaku*, **13**, 627 (1964).

30. F. A. Keidel, "Coulometric analyser for trace quantities of oxygen," *Ind. Eng. Chem.*, **52**, 490 (1960).

31. E. L. Eckfeldt and E. W. Shaffer, "Dissolved oxygen measurement by constant potential derivative coulometry," *Anal. Chem.*, **36**, 2008 (1964).

32. P. A. Hersch, "Galvanic determination of traces of oxygen in gases," *Nature*, **169**, 792 (1952).

33. P. A. Hersch, "Trace monitoring in gases using galvanic systems," *Anal. Chem.*, **32**, 1030 (1960).

34. P. A. Hersch, "Electrochemical gas analyser," U.S. Patent 3,223,608, issued Dec. 14, 1965.

35. J. Janata and H. B. Mark, in *Electroanalytical Chemistry* (A. J. Bard, Ed.), Vol. 3, Marcel Dekker, New York, 1969.

36. A. J. Bard and K. S. V. Santhanam, in *Electroanalytical Chemistry* (A. J. Bard, Ed.), Vol. 4, Marcel Dekker, New York, 1970.

37. P. Delahay, *New Instrumental Methods in Electrochemistry*, Interscience, New York, 1954.

38. H. L. Kies, "Coulometry," *J. Electroanal. Chem.*, **4**, 257 (1962).

39. R. G. Clem, "Coulometry present," *Indust. Res.*, **1973**, 50.

40. F. G. Will, "Electrochemical oxidation of hydrogen on partially immersed platinum electrodes," *J. Electrochem. Soc.*, **110**, 145, 152 (1963).

41. D. R. Kendall, "Improved Hersch cell system for rapid determination of oxygen in gases," *Anal. Chem.*, **43**, 944 (1971).

42. W. J. Baker et al., "A new plant detective—the galvanic cell oxygen analyser," *Ind. Eng. Chem.*, **51**, 727 (1959).

43. R. A. Robinson and R. H. Stokes, *Electrolyte Solutions*, Butterworths, London, 1970.

44. G. Burkert, "Measurement and recording of dissolved oxygen in boiler feed water," *Tekn. Kem. Aik.*, **19**, 198 (1962).

45. R. I. Wright and J. N. Haagen Smith, "Measurement of dissolved oxygen," U.S. Patent, 3,042,495, issued July 3, 1962.

46. J. M. Wright and W. T. Lindsay, "New method for the continuous analysis of dissolved oxygen in water," *Proc. Am. Power Conf.*, **21**, 706 (1959).

47. O. L. Kabanova and E. A. Zalogina, "The determination of trace amounts of dissolved oxygen in water by the use of a thallium column at 32 and 40°C," *Zh. Anal. Khim.*, **20**, 608 (1965).

48. W. Lueck, "Oxygen trace measurements in water by the thallium method," *Energie*, **19**, 203 (1967).

49. W. Lueck, "New service instrument for monitoring the oxygen content of feed water for boilers," *Siemens-Z*, **41**, 527 (1967).

50. W. Lueck, "Oxygen trace measurements in water by the thallium method," *DECHEMA Monogr.*, **61**, 155 (1968).

51. A. N. Doronin, O. L. Kabanova, and S. A. Timofeev, "Thallium column for the determination of oxygen dissolved in water at 80°C," *Zh. Anal. Khim.*, **24**, 274 (1969).

52. W. Lueck, "Industrial measurements of traces of oxygen in light and heavy water by the thallium method," *Messtechnik* (*Brunswick*), **78**, 181, (1970).

53. C. O. Tiller, A. C. Lilly, and B. C. LaRoy "Ionic conduction in LaF_3 thin films," *Phys. Rev. B*, **8**, 4787 (1973).

54. B. C. LaRoy, A. C. Lilly, and C. O. Tiller, "A solid state electrode for reducible gases," *J. Electrochem. Soc.*, **120**, 1668 (1973).

55. A. C. Lilly, B. C. LaRoy, C. O. Tiller, and B. Whiting, "Transport properties of LaF_3 thin films," *J. Electrochem. Soc.*, **120**, 1673 (1973).

56. H. H. Mobius, "Solid electrolyte cell for gas analysis," British Patent, 1,340,084, issued Dec. 5, 1973.

57. D. Priesching, "Device for electrochemical determination of the oxygen content of liquids," German Patent 2,328,921, issued Jan. 24, 1974.

58. L. Z. Zaretskii et al., "Electrochemical detector with automatic restoration of the working surface of a solid indicator electrode," *Zavod. Lab.*, **39**, 786 (1973).

59. B. Fisher and D. S. Tannhauser, "Electrical properties of cobalt monoxide," *J. Chem. Phys.*, **44**, 1663 (1966).

60. N. G. Eror and J. B. Wagner, "Electrical conductivity and thermogravimetric studies of single crystalline cobaltous oxide," *J. Phys. Chem. Solids*, **29**, 1597 (1968).

61. I. Bransky and J. M. Wimmer, "The high temperature defect structure of CoO," *J. Phys. Chem. Solids*, **33**, 801 (1972).

62. J. M. Wimmer, R. N. Blumenthal, and I. Bransky, "Chemical diffusion in CoO," *J. Phys. Chem. Solids*, **36**, 269 (1975).

63. C. R. McIlwrick and C. S. G. Phillips, "The removal of oxygen from gas streams," *J. Phys. E: Sci. Instr.*, **6**, 1208 (1973).

64. E. M. Logothetis et al., "Oxygen sensors using CoO ceramics," *Appl. Phys. Letters*, **26**, 209 (1975).

65. A. Duquesnoy and F. Marion, "On the variation of the conductivity of CoO, NiO and MnO as a function of the equilibrium partial pressure of oxygen at high temperatures," *Compt. Rend.*, **256**, 2862 (1963).

66. F. Marion, A. Duquesnoy, and P. Dherbomez, "On the efficiency of the processes of deoxygenation of non-combustible gases," *Bull. Soc. Chim. France*, **1964**, 1290.

67. H. L. Tuller and A. S. Nowick, "Doped ceria as a solid oxide electrolyte," *J. Electrochem. Soc.*, **122**, 255 (1975).

68. H. Degn and J. McK. Nobbs, "Solid state oxygen meter using zinc oxide," *Appl. Phys. Letters*, **26**, 526 (1975).

69. K. H. Mancy, "Analytical problems in water pollution control," in *Analytical Chemistry: Key to Progress on National Problems*, NBS Special Publication 351, 1972.

70. F. Todt, *Electrochemical Oxygen Measurements*, W. deGruyter and Company, Berlin, 1958.

71. J. H. Karchmer, "Polarographic determination of dissolved oxygen—study of drop time with rapid dropping electrode," *Anal. Chem.*, **31**, 502 (1959).

72. C. P. Tyler and J. H. Karchmer, "Portable analyser for determination of dissolved oxygen in water—application of rapid dropping mercury electrodes," *Anal. Chem.*, **31**, 499 (1959).

73. R. Briggs, G. V. Dyke, and G. Knowles, "Use of the wide bore dropping mercury

electrode for long period recording of concentration of dissolved oxygen," *Analyst*, **83**, 304 (1958).

74. M. C. Rand and M. Heukelekian, "Determination of dissolved oxygen in industrial wastes by the Winkler and polarographic methods," *Sew. & Ind. Wastes*, **23**, 1141 (1951).

75. F. Todt, "Application of the electrochemical determination of oxygen," *Angew. Chem.*, **9**, 266 (1955).

76. H. Ambuhl, "Practical application of the electrochemical determination of oxygen in water," *Schweiz. Z. Hydrol.*, **22**, 23 (1960).

77. M. L. Hitchman, unpublished results.

78. H. A. Laitinen, T. Higuchi, and M. Czuha, "Potentiometric determination of oxygen using the dropping mercury electrode," *J. Am. Chem. Soc.*, **70**, 561 (1948).

79. F. R. Smith and P. Delahay, "Coulostatic study of residual oxygen in electrochemical investigations," *J. Electroanal. Chem.*, **10**, 435 (1965).

80. A. A. Wood, "Operation of a wide bore polarograph, continuous dissolved oxygen recorder," *Effl. and Water Treat. J.*, **1963**, 23.

81. J. Tenygl and B. Fleet, "A coulometric oxygen analyser based on a porous catalytic electrode with vacuum controlled active surface area," *Coll. Czech. Chem. Comm.*, **38**, 1714 (1973).

82. G. Eden et al., "Miniaturised electrode for on-line pO_2 measurements," *IEEE Trans. Biomed. Eng.*, **22**, 275 (1975).

83. L. Heyne, "Some properties and applications of zirconia based solid electrolyte cells," in *Measurement of Oxygen* (H. Degn, I. Balslev, and R. Brook, Eds.), Elsevier Scientific Publishing Company, Amsterdam, 1976.

84. M. Kleitz and J. Fouletier, "Technological improvements and accuracy of potentiostatic measurements with zirconia gauges," in *Measurement of Oxygen* (H. Degn, I. Balslev, and R. Brook, Eds.), Elsevier Scientific Publishing Company, Amsterdam, 1976.

85. W. J. Fleming, "Physical principles governing non-ideal behavior of the zirconia oxygen sensor," *J. Electrochem. Soc.*, **124**, 21 (1977).

86. I. Fatt, *Polarographic Oxygen Sensors*, CRC Press, Cleveland, 1976.

OTHER METHODS OF MEASUREMENT—NONELECTROCHEMICAL

"And thick and fast they came at last, and more and more and more."

Lewis Carroll, *The Walrus and the Carpenter*

8.1. CHEMICAL METHODS

8.1.1. WINKLER METHOD AND ITS MODIFICATIONS

Probably the oldest standard method of analysis for dissolved oxygen is that first reported by Winkler in 1888 (1). And although the method has been improved by variations in equipment and technique, and has been aided by instrumentation, it is still the basis for the majority of titrimetric analyses. The method is based on the quantitative oxidation of Mn(II) to Mn(III) in alkaline solution with the subsequent oxidation of I^- by the Mn(III) in acid solution. The iodine which is liberated is titrated with thiosulphate in the normal way. The equations for the various steps are

$$Mn^{2+} + 2OH^- \rightarrow Mn(OH)_2\downarrow$$
$$2Mn(OH)_2 + \tfrac{1}{2}O_2 + H_2O \rightarrow 2Mn(OH)_3$$
$$2Mn(OH)_3 + 6H^+ + 3I^- \rightarrow 2Mn^{2+} + I_3^- + 6H_2O$$
$$I_3^- \rightarrow I_2 + I^-$$
$$I_2 + 2S_2O_3^{2-} \rightarrow 2I^- + S_4O_6^{2-}$$

The method depends on the careful control of pH and $[I^-]$, and failure to do this can lead to significant errors (2, 14). If strict control is maintained the precision of the method is better than ± 0.1 mg liter^{-1} of dissolved oxygen (3) [Section 5.2.3].

Because various ions and compounds cause interference with the Winkler method a number of modifications have been introduced to correct for these interferences. Table 8.1 summarizes the standard modifications as recommended by the American Public Health Association (APHA) (3) and full details, together with references to original articles, can be found in this publication for methods for the examination of water and wastewater.

TABLE 8.1 Standard Modifications of the Winkler Method (3)

Name of modification	When used	Reagents	Procedure	Notes
A. Azide [Alsterberg's method (4)]	In presence of >0.1 mg liter^{-1} nitrite and <1 mg liter^{-1} Fe^{2+}, e.g. in biologically treated effluent such as sewage	2.15M MnSO$_4$: 0.9 moles NaI (or KI) and 0.15 moles NaN$_3$ (or KN$_3$) added to 1 liter 12.5M NaOH (or KOH); conc. H$_2$SO$_4$; 0.5% starch solution: 0.025M Na$_2$S$_2$O$_3$·5H$_2$O	Standardize Na$_2$S$_2$O$_3$ with 2.1 mM KH(IO$_3$)$_2$ or 4.2 mM K$_2$Cr$_2$O$_7$. To 300 ml DO sample add 2 ml MnSO$_4$ + 2 ml alkali-iodide-azide solutions. Allow Mn(OH)$_3$ floc to settle. Add 2 ml conc. H$_2$SO$_4$. Shake to get uniform I$_2$ distribution. Take 203 ml of solution and titrate with Na$_2$S$_2$O$_3$ with starch for end point. Calculate DO concentration: 1 ml Na$_2$S$_2$O$_3$ ≡ 1 mg liter^{-1} O$_2$	Reducing and oxidizing agents must be absent, but addition of 1 ml of 4.25M KF before acidifying makes method applicable to solution with <100–200 mg liter^{-1} Fe^{3+}
B. Permanganate [Rideal—Stewart method (5)]	In presence of Fe^{2+} (and of Fe^{3+} if 1 ml KF is added before acidification) and reducing organic matter	Same as for method A plus 0.04M KMnO$_4$ and 0.1M potassium oxalate, K$_2$C$_2$O$_4$	To 300 ml DO sample add 0.7 ml conc. H$_2$SO$_4$, 1 ml KMnO$_4$, and 1 ml KF. Mix. If permanganate color fades add more KMnO$_4$. Add 0.5–1 ml K$_2$C$_2$O$_4$, mix, and stand in dark. Permanganate color should fade in 2–10 min. Avoid excess K$_2$C$_2$O$_4$ or otherwise low DO values obtained.	Method is ineffective for oxidation of SO$_3^{2-}$, S$_2$O$_3^{2-}$, polythionate, or organic matter—large errors in presence of these

161

TABLE 8.1 (*cont.*)

Name of modification	When used	Reagents	Procedure	Notes
			Add 2 ml $MnSO_4$ + 3 ml alkali-iodide-azide. Allow precipitate to settle. Add 2 ml conc. H_2SO_4. Take 205 ml of solution and titrate with $Na_2S_2O_3$	
C. Alkali-hypochlorite	In presence of SO_3^{2-}, $S_2O_3^{2-}$, and polythionate	Same as for method A plus $2M$ NaOCl, $1M$ NaI (or KI), $0.05M$ $Na_2S_2O_3$, $(1+9)H_2SO_4$.	To 300 ml DO sample add 0.2 ml NaOCl to oxidize SO_3^{2-}. Add 1 ml I$^-$ and 1 ml $(1+9)$ H_2SO_4. Neutralize liberated I_2 with Na_2SO_3 using starch indicator. Add $KH(IO_3)_2$ to just restore blue color. Continue as in method A. Take volume equivalent to 200 ml of original sample and titrate with $Na_2S_2O_3$	

D. Alum-flocculation	In presence of suspended solids which could consume I_2	Same as for method A plus $0.2M$ $KAl(SO_4)_2 \cdot 12H_2O$ and conc. NH_4OH	To 500 ml DO sample add 10 ml alum solution + 1–2 ml conc. NH_4OH. Allow precipitate to settle. Siphon off 300 ml of supernatant liquor and treat as in method A	Accuracy of method is low (59)
E. Copper sulfate—sulfamic acid flocculation	In presence of biological flocs with high O_2 consumption, e.g. activated sludge	Same as for method A plus inhibitor solution: $0.33M$ NH_2SO_2OH + $0.2M$ $CuSO_4 \cdot 5H_2O$ + CH_3COOH	To 10 ml inhibitor solution add 1 liter DO sample. Allow suspended solids to settle. Siphon off 300 ml of supernatant liquor and treat as in Method A	
F. "Short"	Organic substances which are readily oxidized in strong alkali or by the I_2 in acid solution	Same as for method A	Same as for method A except 2 ml conc. H_2SO_4 added before $Mn(OH)_2$ floc settles	
G. High DO or organic content [Pomeroy–Kirschman method (6)]	> 15 mg liter^{-1} DO or high organic content, e.g. domestic sewage	Same as for method A except alkali-iodide-azide solution is $6M$ in NaI and $10M$ in NaOH	Same as for method A	KI cannot be used because of limited solubility. Method applies for DO up to 40 mg liter^{-1}

163

Many other modifications have been suggested for specific interferences and also to simplify or improve the basic method (59, Ch. 3); one interesting modification allows analysis of small volume samples—1 to 2 ml. The biennial review of water analysis published by the American Chemical Society (16) and the annual review of water treatment and analysis by the Water Pollution Control Federation (17) provide an excellent starting point for finding out more about these many methods. In general, however, very few new methods have been validated to the extent that the APHA methods have.

In any method for dissolved oxygen analysis which involves taking a sample of the water it is, of course, essential that atmospheric oxygen is not entrained or dissolved in the sample. Errors in sampling have been discussed by Montgomery and Cockburn (7), and reference 3 discusses special precautions to avoid oxygen contamination of the sample. In general, narrow-mouthed sample bottles are completely filled and stoppered while immersed—which can be done either manually or automatically—or special samplers may be used. A very simple flask into which a water sample can be drawn is shown in Fig. 8.1a, and the Kemmerer water sampler (3) in Fig. 8.1b. The latter method of sampling is especially recommended for measurements in natural waters at depths greater than a meter or so (3); the water is transferred from the sampler to an analysis bottle via a drain hole at the bottom of the sampler. Sampling techniques for waters are clearly not peculiar to dissolved oxygen analysis and the general questions of sampling methods, sampling sites, sampling times, and sampling frequencies have been examined in various places (3, 60–62).

Samples once collected should be determined with as little delay as possible, especially if there is some biological activity which is consuming oxygen; the addition of 0.7 ml concentrated sulfuric acid and 1 ml of $0.3M$ sodium azide will arrest such activity for a number of hours (3). Once the determination is started the addition of all solutions prior to the acidification of the sample should be made well below the surface of the sample. Sample addition need not, of course, be restricted simply to the manual use of pipettes and burettes, and automatic titrations as well as expendable ampoules containing the Winkler reagents have been suggested (52); the latter technique is particularly useful for analyses in the field. Whichever method of addition is used though, it is clear that some of the sample from the bottle will be displaced, and it is for this reason that a volume correction for the sample loss is taken for the titration; for example, 203 ml instead of 200 ml for the azide modification—Table 8.1.

The greatest error in Winkler-based determinations arises in the last step of the analysis, the iodine titration. First, it is well known that the I^- ion is highly susceptible to aerial oxidation, especially in the presence of catalytic metal ions, and so high readings can result if precautions are not

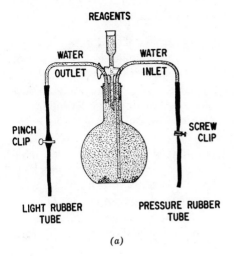

REAGENTS

WATER
OUTLET

WATER
INLET

PINCH
CLIP

SCREW
CLIP

LIGHT RUBBER
TUBE

PRESSURE RUBBER
TUBE

(a)

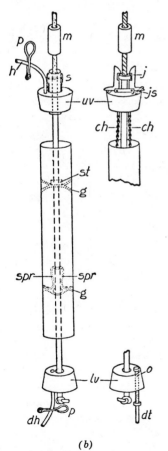

p *m*

h

s

uv

ch *ch*

st

g

spr *spr*

g

lv

dh *p*

(b)

m

j

js

o

dt

Left: view of complete sampler with valves open. Top right: another type of construction of upper valve and tripping device. Bottom right: another type of construction of lower valve and drain tube. Key: *ch*—chain which anchors upper valve to upper interior guide; *dh*—rubber drain tube; *dt*—brass drain tube; *g*—interior guide fastened to inner surface of body of sampler; *h*—rubber tube; *j*—jaw of release; *js*—jaw spring; *lv*—lower valve; *m*—messenger; *o*—opening into interior of drain tube; *p*—pinch cock; *s*—upper release spring operating on horizontal pin, one end of which fits into groove on central rod; *spr*—spring fastened to lower internal guide and operating in groove on central rod to provide lower release; *st*—stop on central rod; *uv*—upper valve.

Fig. 8.1 Devices for collecting water samples for dissolved oxygen analysis. (*a*) Flask for Winkler analysis. (*b*) Modified Kemmerer sampler (3).

165

taken to prevent such oxidation occurring during the titration. Second, and perhaps more important, iodine is rather volatile and loss of the vapor can lead to low readings. Montgomery, Thom, and Cockburn (2) have examined this question in considerable detail and conclude that in fact loss of iodine vapor in the standard modifications of the Winkler method is appreciable, especially if the titrations are carried out slowly. In order to reduce iodine losses they recommend that the alkaline iodide reagent of Pomeroy and Kirschmann (6) be used instead of the standard solution. This reagent is $6M$ in NaI and $10M$ in NaOH and so is identical to that of modification G of the APHA—Table 8.1. However, this solution is now suggested for use with all concentrations of dissolved oxygen and not just for $[O_2] > 15$ mg liter^{-1}. Figure 8.2 compares results of Montgomery and others using a high $[I^-]$ with those obtained by various other workers using different methods.

Two things are immediately apparent from Fig. 8.2: first, the good agreement between the modified Winkler method and the careful analyses by Klots and Benson, and by Fox, both using gasometric methods; second, the large deviations of the results of Truesdale and others, who used the standard Winkler method, from those obtained with the modified method. Both these observations confirm the efficacy of the modified Winkler method in overcoming the errors arising from loss of iodine vapor. The results of Carlson, Elmore and Hayes, and Winkler himself show, however, that with some care the method using the standard alkali-iodide solution can yield reasonable results, but it is probably worth ensuring accurate analyses by using the procedure recommended by Montgomery and co-workers. Carritt and Carpenter (8) have confirmed that the loss of iodine through volatilization is the main source of error in the Winkler methods; they have also shown that the photochemical oxidation of I^- is an additional problem, thus emphasizing the need for as little delay as possible in carrying out the thiosulfate titration. An analysis of sources of error in the Winkler method has been given by Green (27), who has also suggested an improved iodine determination flask for the method (56).

Since the last step in any of the standard modified Winkler methods is the weak link in the chain, a number of alternatives have been suggested. The simplest of these is just to omit the final titration and to measure the amount of coloration produced by the triiodide ion. This can be done using standard colored discs, and a comparison is made either directly with the aqueous triiodide solution which is obtained on acidification (9, method 2), or the iodine is extracted into carbon tetrachloride and then the comparison is made (9, method 3); with care and practice accurate results can be obtained. Alternatively, the I_3^- ion can be determined spectrophotometrically by measuring the uv absorption at 352 nm (11, 64).

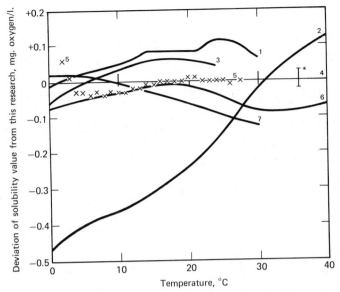

Fig. 8.2 Comparison of oxygen solubility values in pure water obtained by various groups of workers (2).

1. G. B. Whipple and M. C. Whipple, *J. Am. Chem. Soc.*, **33**, 362 (1911)
2. G. A. Truesdale, A. L. Downing, and G. F. Lowden, *J. Appl. Chem.*, **5**, 53 (1955)
3. T. Carlson, *Akad. Afh. Stockholm*, **1912**
4. H. A. C. Montgomery, N. S. Thom, and A. Cockburn, *J. Appl. Chem.*, **14**, 280 (1964)
5. C. E. Klots and B. B. Benson, *J. Mar. Res.*, **21**, 48 (1963)
6. L. W. Winkler, *Ber. Deutsch. Chem. Ges.*, **24**, 3602 (1891)
7. H. L. Elmore and T. W. Hayes, *Proc. Am. Soc. Civ. Engrs.*, *J. San. Eng. Div.*, **86** (SA4), 41 (1960)

* Root mean square deviation of work of reference 4. [Reproduced by permission of the authors (2).]

The detection of the end point in the final titration can be improved and this leads to increased sensitivity, accuracy, and precision. Several methods exist for doing this, all based on electrochemical techniques. The first method involves a potentiometric titration (12, Ch. 3; 13). A platinum wire electrode monitors the potential of the solution with respect to a suitable reference electrode (e.g. saturated calomel electrode) as the titration proceeds. Initially the potential is determined by the I_2/I^- couple according to eq. (3.9), but as the I_2 is removed the potential changes, slowly at first and then more rapidly as the equivalence point is approached. At the equivalence point there are, as the name implies, an equal number of equivalents of I_2 and $S_2O_3^{2-}$, and then beyond the equivalence point there is so little I_2 left that the potential

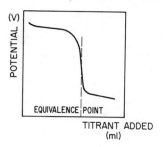

Fig. 8.3   Potentiometric titration curve.

monitored is determined by the $S_2O_3^{2-}/S_4O_6^{2-}$ couple. The equivalence or stoichiometric end point is readily detected as an inflexion point in the titration curve—Fig. 8.3. A second method uses the limiting current at, say, a rotating platinum wire electrode for iodine reduction (14). Plotting the limiting current as a function of titrant added clearly shows the end point by an abrupt change in the slope of the curve; in this case the slope will change from being negative to zero as no further iodine is available for reduction and no other species is capable of being reduced—Fig. 8.4. This type of end point detector is known as an amperometric titration (12, Ch. 5).

A rather simpler, although just as accurate, amperometric technique is the "dead stop" end point (15, Ch. 10). Here, instead of having an electrode with a well-characterized limiting current, two platinum ·electrodes are immersed in the titration cell, the solution is stirred, a constant voltage of the order of 10–100 mV is applied to the electrodes, and as the titration is carried out the current is monitored—Fig. 8.5 shows a suitable circuit to do this. It should be noted that the potential difference between the two electrodes is controlled and not the potential of one electrode with respect to a reference. The titration curve obtained is the same as that found with amperometric titrations, and again the end point is indicated by the almost complete disappearance of the current flowing in the cell—Fig. 8.6. The explanation of the shape of the curve is also essentially the same as that for

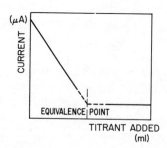

Fig. 8.4   Amperometric titration curve.

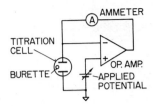

Fig. 8.5 Circuit for "dead stop" end point.

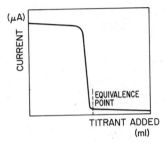

Fig. 8.6 Titration curve for "dead stop" end point.

the amperometric titration. At the start of the titration both components of the reversible I_2/I^- couple are present and appreciable current flows through the cell as I_2 is reduced at the cathode and I^- oxidized at the anode. With the removal of I_2 during the titration the current falls as there is no readily reducible species to replace it—the $S_2O_3^{2-}/S_4O_6^{2-}$ couple is very irreversible—and at the end point when no I_2 is left the cell current comes to a "dead stop." Using a sensitive current measuring device the end point can be determined to ± 0.01 μg of I_2 in a 100 ml sample, and so this method provides a relatively simple means of improving the accuracy of the final stage of the Winkler method.

However, the "dead stop" end point titration still suffers from the drawbacks of possible errors being introduced by loss of iodine vapor. This problem is overcome by the generation of iodine *in situ* using constant current coulometry. The procedure for the Winkler method is altered slightly in that on acidification excess standard $Na_2S_2O_3$ solution is added, so consuming immediately all the iodine liberated by the acid. The electrolytic generation of iodine is then started, and provided the solution is well stirred the back titration of the thiosulfate can be followed by the "dead stop" end point method; the curve will, of course, be the mirror image of Fig. 8.6 since the reversible couple is being formed. Figure 8.7 illustrates a typical circuit for constant current coulometry. The calculation of the dissolved oxygen concentrations is readily made using eq. (3.1).

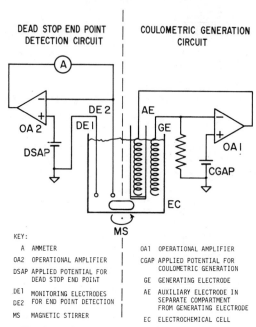

DEAD STOP END POINT
DETECTION CIRCUIT

COULOMETRIC GENERATION
CIRCUIT

KEY:

A   AMMETER

OA2   OPERATIONAL AMPLIFIER

DSAP   APPLIED POTENTIAL FOR
       DEAD STOP END POINT

DE1   MONITORING ELECTRODES
DE2   FOR END POINT DETECTION

MS   MAGNETIC STIRRER

OA1   OPERATIONAL AMPLIFIER

CGAP   APPLIED POTENTIAL FOR
       COULOMETRIC GENERATION

GE   GENERATING ELECTRODE

AE   AUXILIARY ELECTRODE IN
       SEPARATE COMPARTMENT
       FROM GENERATING ELECTRODE

EC   ELECTROCHEMICAL CELL

Fig. 8.7   Circuit for constant current coulometry.

8.1.2. COLORIMETRIC METHODS

A considerable number of colorimetric procedures have been developed over the years for dissolved oxygen. Hamlin and Lambert (10) cite 12 references to such methods, and several reviews and comparisons of the various methods have been made (16, 18–20, 59). The attraction of colorimetric methods is that they provide neat, quick, and quite accurate procedures for routine measurements of dissolved oxygen, especially for sub-ppm concentrations. Also it is generally possible to adapt the methods for field use by technicians.

To describe here or even mention all the colorimetric methods that have been suggested in the literature would be too lengthy, but we can illustrate the general principles behind all the methods by outlining one particular method; for information about other specific methods reference can be made to the reviews already mentioned, especially to the more accessible articles in (16).

The method we choose as an illustration is the indigo-carmine test (9, method 1; 63). This test was originated by Buchoff, Ingber, and Brady (21) as a determination for low, dissolved oxygen concentrations (<1 ppm) as found, for example, in boiler feed waters. Reduced indigo-carmine (the leuco form) is a bright yellow-green color, and on oxidation it changes to orange, through red, purple, and blue to finally become an intense blue-green. The leuco reagent can be prepared by glucose reduction in alkaline solution of the oxidized form and once it has been obtained it can be mixed with the water sample, taking the usual precautions to avoid contamination by atmospheric oxygen. After a few minutes mixing, the oxidation reaction is complete and the fully developed color can be compared with standard samples or with permanent glass standards (9); alternatively spectrophoto-metric monitoring can be employed. Interference with the test can arise from Fe^{2+} and this ion, together with other intolerable ions (especially oxidizing ions), can be most readily removed with a mixed-bed ion-exchange resin in series with the sampling line (21); nonionic oxidizing agents are more difficult to cope with.

Other colorimetric methods may differ in the preparation of the reduced form of the colored compound [e.g. some can be photoreduced (10)], in the wavelength used for spectrophotometric monitoring, and in the specific interferences that occur, but in all cases the overall procedure remains the same. The main disadvantages of colorimetric methods are that they can only be used in relatively clean samples, and that they are not suitable for *in situ* or continuous monitoring. Nevertheless, some of the tests have been shown to be sufficiently reliable and accurate to form the basis for standard methods.

8.1.3. OTHER CHEMICAL METHODS

In addition to adaptations of the Winkler method [i.e. where the initial step is the "fixing" of the oxygen by oxidation of Mn^{2+}, but the subsequent analysis does not involve iodine generation, e.g. (22–24)] and colorimetric methods such as we have just described, there are a few further methods for dissolved oxygen analysis. These methods are titrimetric methods which do not use Mn(II) oxidation as the starting point. One such method (25) uses the transformation of tetramine copper(I)—chloride complex, which is formed in a solution of NH_4OH/NH_4Cl containing Cu powder, into the Cu(II) complex by the addition of dissolved oxygen. The resulting solution is heated to boiling and titrated with EDTA to a yellowish-green color with 1-(2-pyridylazo)-2 naphthol as indicator. Another method (26) replaces the Mn(II) of the Winkler method by Ce(III), which dissolved oxygen oxidizes to Ce(IV) in alkaline solution. Acidification of the ceric hydroxide produces a very stable solution of $Ce(SO_4)_2$ and this can be titrated with the primary standard, arsenious oxide. This last method can also be treated as a colori-metric method since the $Ce(SO_4)_2$ is yellow and the color does not fade, except in the presence of reducing agents. The Miller method (54) involves the titration of dissolved oxygen with a ferrous solution, the end point being detected with phenosafranine redox indicator. This method, like the other methods mentioned here, avoids the problem of loss of iodine and in addition has the advantages of speed and simplicity. Unfortunately it is not, in general, very precise, due mainly to the end point being highly dependent on pH. Carefully standardized conditions must therefore be used to obtain accurate and reproducible results, and the precautions necessary to achieve optimum results with the method have been described (55).

An interesting method described recently (57) allows precise ($\pm 0.3\%$) and rapid (~ 3 min) analyses of small samples (~ 1 ml) to be made. An acid solution of nitrite and iodide react in the presence of oxygen, generating iodine. By coulometric reduction of the iodine the amount of oxygen can be determined, and the results agree well with the average values given in Table 2.3.

Further chemical methods are mentioned in reference 16, but, like many modifications to the Winkler method, few have been tried extensively enough to warrant an unreserved recommendation for general use.

8.1.4. SUMMARY OF CHEMICAL METHODS

Classical methods of analysis suffer from a number of drawbacks. They are rather slow even when most of the steps, from sampling through addition of reagent to end point determination, are automated. They cannot be used for *in situ* measurements, nor can they give continuous monitoring. They are often subject to interference from substances commonly found in water samples. They can yield inaccurate and misleading results due to lack of care and precision in methodology. Yet in spite of these disadvantages, chemical methods, and the Winkler method in particular, provide the most readily accessible means of obtaining standard analyses of dissolved oxygen samples.

The APHA (3) quotes the Winkler method as allowing the dissolved oxygen content of water to be determined with a precision, expressed as a standard deviation, of 0.043 mg liter^{-1}. However, in the presence of interferences, even with the proper modification, the standard deviation may be as high as 0.1 mg liter^{-1}. Taking air-saturated water at 25°C, the more precise determination corresponds to an error of about 0.5%, which is of the same order as the average deviation found in seven independently determined values for oxygen solubility at the same temperature (28). Since, therefore, oxygen solubilities are not generally known much more precisely than $\pm 0.5\%$ the precision of the Winkler method would appear to be adequate. The precision of other chemical methods and colorimetric methods (9, 16) is invariably less than that for the Winkler method, and if care is not taken errors can often be as high as 10% or more.

8.2. GASEOUS PHASE MEASUREMENTS

Since we are interested in analysis of dissolved oxygen, monitoring in the gaseous phase obviously requires the aqueous sample to be completely degassed. The main reason for going to such lengths is to overcome the difficulties associated with measurements made in the presence of interfering substances. The MPOD is much less susceptible to interference than most of the chemical methods or the other electrochemical methods. But, even so, interferences do occur [Section 5.2.6] with the additional problem of poisoning of the electrodes by polluting gases [Section 5.3.5]. Therefore the only effective way to deal with all these problems is to remove the oxygen from solution and to purify it from contaminating gases before making the measurement.

8.2.1. GAS EXTRACTION METHODS

Clearly, for gas phase analysis complete or reproducible removal of the oxygen from the aqueous sample is important. Battino and Clever (28) give a useful summary of gas extraction methods as well as of criteria for complete degassing. Of the various methods that are available the most practicable for routine analysis is that based on the technique of bubbling an inert gas through the sample to strip it of oxygen. Figure 8.8 shows a gas exchange unit schematically. Water, taken in from the source, is forced with the carrier gas through a nozzle in the aspirator at several atmospheres pressure. Provided that careful control of water and gas flow rates is maintained, then rather efficient transfer of oxygen from the aqueous phase is achieved (29, Ch. XIII). On leaving the exchange unit the carrier gas and the oxygen pass into the analyzer and then are either vented or recirculated through the exchange unit in order to achieve equilibrium conditions.

A very simple but efficient gas stripper has been described by Williams and Miller (30). The principle of their device is shown in Fig. 8.9. As the Mylar discs rotate through the liquid phase a thin film of the liquid is spread over their surfaces and exposed to the gas phase giving rise to a very rapid rate of exchange. Using flow rates up to 100 ml min^{-1} and He/H$_2$O ratios of 10:1 through to 1:2 they found 100% gas removal. A small unit such as this (10 × 5 cm) could readily be used in place of the aspirator of Fig. 8.8.

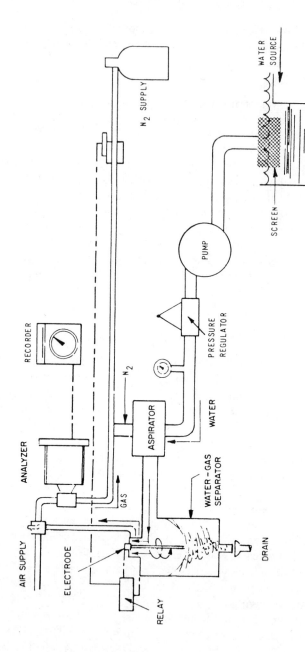

Fig. 8.8 Gas exchange unit (29). (Reproduced by permission of Ann Arbor Science Publishers Inc.)

174

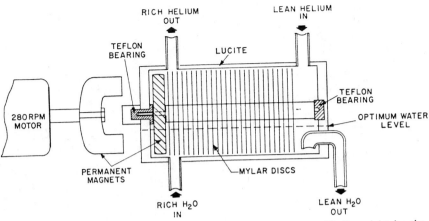

Fig. 8.9 Gas stripper. (Reproduced by permission from reference 30. Copyright by the American Chemical Society.)

8.2.2.   GAS MONITORING METHODS

Once the dissolved oxygen has been stripped from the sample under test it could be redissolved in clean water and then monitored chemically or electrochemically. Redissolution, however, only introduces complications and generally analysis is made of the extracted gas. A MPOD could be used for this purpose, and its use here is very similar to that which we have already covered extensively for direct dissolved oxygen measurements. Two points are worth noting though. One is that because gaseous diffusion is a much more rapid process than diffusion in solution, depletion outside the membrane is not really a problem, and no special steps need to be taken to ensure adequate forced convection [cf. Section 5.3.1]. The other point is that drying out of the cell will occur more rapidly than when a MPOD is used in an aqueous environment and attention must be paid to this [Section 5.1.3].

Manometric—Volumetric Methods.   Although the Winkler method is the oldest standard method of dissolved oxygen analysis, measurements of volumes of gases, together with solubility determinations by absorption methods, predate the chemical method by 30 years or more. Such methods are capable of a high degree of accuracy and are valuable for obtaining standard values (35, 36, 53). The review of Battino and Clever (28) on the solubility of gases in liquids summarizes the various designs of volumetric apparatus that have been used. The manometric method of Van Slyke (68),

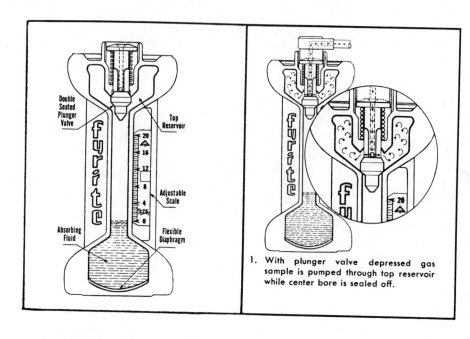

1. With plunger valve depressed gas sample is pumped through top reservoir while center bore is sealed off.

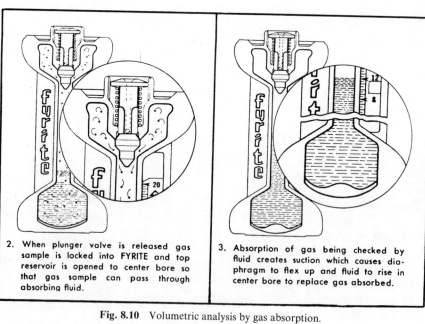

2. When plunger valve is released gas sample is locked into FYRITE and top reservoir is opened to center bore so that gas sample can pass through absorbing fluid.

3. Absorption of gas being checked by fluid creates suction which causes diaphragm to flex up and fluid to rise in center bore to replace gas absorbed.

Fig. 8.10 Volumetric analysis by gas absorption.

often still used by physiologists for blood analysis, is described in considerable detail in the author's original paper, and it has recently been reevaluated by Zander and Euler (69). However, from a practical standpoint neither manometric nor volumetric methods are generally suitable for routine monitoring unless considerable simplifications are made in the techniques. One such simplification is the microgasometric method of Scholander and co-workers (65). This method strips off the dissolved gases in a specially designed pipette after which the gases, in bubble form, are first treated with alkali to remove carbon dioxide and then with alkaline pyrogallol to absorb the oxygen. The absorption produces a change in the gas bubble volume which is equivalent to the amount of oxygen present. The same principle is used in a commercial analyzer—Fig. 8.10. Other gases can also be analyzed by the same technique, but, of course, continuous monitoring can never be possible.

Mass Spectrometry. The principles of mass spectrometry can be quickly gleaned from any modern physical chemistry text (e.g. 31, Ch. 11), and Fig. 8.11 illustrates these principles. Various problems experienced with the application of mass spectrometry to the analysis of oxygen and the methods adopted to overcoming them have been outlined by Nobbs (70). However, as a means of solely determining dissolved oxygen concentrations mass spectrometry is an unnecessarily complicated method, although water samples have been analyzed in this way (32, 33). The real usefulness of the method comes when isotopic analysis of oxygen is required (34; 67, p. 2).

Thermal Conductivity. Diffusion, we have seen, is the transport of mass across a concentration gradient [Section 3.3.3]. The transfer of heat as kinetic energy across a temperature, or kinetic energy, gradient is an analogous process and is known as *thermal conductivity*. It can be defined as the amount of heat transferred in unit time across surfaces of unit area, through unit distance, and for a unit temperature difference between the surfaces; the units are, in general, energy distance^{-1} time^{-1} degree^{-1}, which in SI units [Appendix A] become J m^{-1} s^{-1} K^{-1}.

In order to analyze oxygen using thermal conductivity the gas is passed, with the carrier gas, into a cell containing a metal wire with a high temperature coefficient of resistance (e.g. platinum). The wire forms one arm of a Wheatstone bridge network and when a current is passed through the bridge the wire heats up. The temperature that the wire attains depends not only on the current flowing, but on the thermal conductivity of the surrounding gas as well. As the composition of the surrounding gas changes with the oxygen extracted from the aqueous sample its thermal conductivity also changes. This change gives rise to an imbalance in the bridge network and this imbalance is measured using a galvanometer or a potentiometer

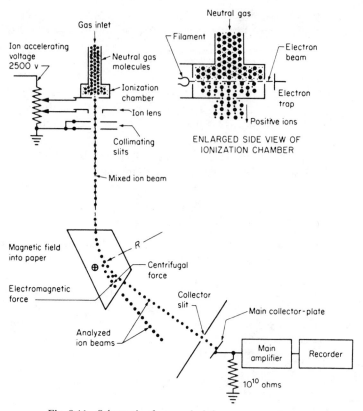

Fig. 8.11 Schematic of gas analysis by mass spectrometry.

which can readily be calibrated in terms of the dissolved oxygen in the test solution. Figure 8.12 illustrates the principle of the technique (29).

A modification of the method we have just described either uses hydrogen as the carrier gas or the exchanged oxygen is mixed with hydrogen (29). The gases are then burnt in the presence of a noble metal catalyst filament which again is part of a resistance network. As before the filament temperature and resistance are related to the oxygen content of the gas sample and hence to the dissolved oxygen concentration.

Paramagnetic Method.   Just as molecules can either have a permanent electric dipole moment or have one induced by an electric field (31, Ch. 14), so they can have a permanent magnetic moment or they can have a moment induced by a magnetic field. Thus all substances, to a greater or lesser degree, interact with magnetic fields and this interaction can take one of

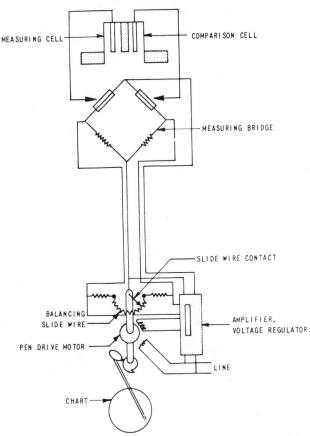

MEASURING CELL

COMPARISON CELL

MEASURING BRIDGE

SLIDE WIRE CONTACT

BALANCING
SLIDE WIRE

AMPLIFIER,
VOLTAGE REGULATOR

PEN DRIVE MOTOR

LINE

CHART

Fig. 8.12 Schematic of oxygen analysis by measurement of thermal conductivity (29). (Reproduced by permission of Ann Arbor Science Publishers Inc.)

two forms. If the magnetic lines of force are drawn into a material placed in a field so that the field in the material is greater than in free space, then the substance is said to be *paramagnetic*—Fig. 8.13*b*. On the other hand, if the lines of force are pushed out of the material so that the field in it is less than in free space the substance is *diamagnetic*—Fig. 8.13*c*. These two types of magnetic materials are readily distinguished experimentally using a magnetic balance. A specimen of the material is hung from a balance beam so that it is suspended partly inside and partly outside the region between the poles of an electromagnet. When the magnet is switched on and the field applied a paramagnetic material is drawn into the field while a diamagnetic sample is pushed out. The displacement of the sample, or the

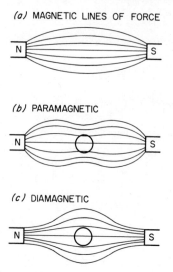

Fig. 8.13　Types of magnetic material.

force needed to restore the balance beam to its original balance point, is a measure of the magnetic susceptibility of the material.

Oxygen is one of the few gases that shows paramagnetic characteristics [Fig. 8.14] and so it can be analyzed by a similar method to that which we have just outlined. Since we are dealing with a gas though, a small modification has to be made. In place of the suspended magnetic sample a small

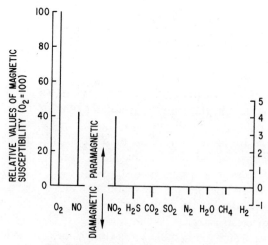

Fig. 8.14　Magnetic properties of gases.

glass dumbbell on a taut, quartz fiber is used. When no oxygen is present the torque of the fiber and the magnetic field are balanced and the dumbbell remains stationary; this "zeroing" of the instrument is usually done with the carrier gas only flowing. On introducing a mixture of carrier gas and oxygen the magnetic field is changed by virtue of the paramagnetic properties of the oxygen, and the dumbbell rotates. The degree of rotation is proportional to the change in the magnetic field and this in turn is proportional to the concentration of oxygen in the sample. Figure 8.15 illustrates the principle of the technique.

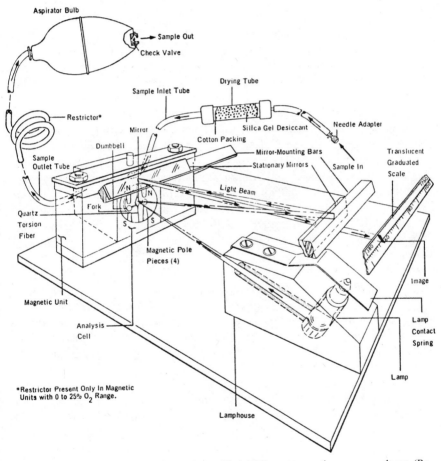

Fig. 8.15 Functional diagram of a Beckman Model D2 paramagnetic oxygen analyzer. (Reproduced by permission of Beckman Instruments Inc.)

Another method combines the use of the paramagnetic properties of oxygen with the thermal conductivity method of analysis. A hot spiral wire is held in a permanent magnetic field. Oxygen gas enters the region of the field and is heated by the wire which causes the gas to lose its paramagnetism. Cooler gas is attracted into the field to replace that which has been heated and this results in the oxygen circulating around and cooling the wire. The resistance of the wire changes and the change is monitored and related to the dissolved oxygen concentration in the way already described. Analysis by use of the paramagnetism of oxygen has been described in the literature (37; 38; 67, p. 55), and one feature which is attractive about the method is its rapid response to changes in p_{O_2}—90% in about 1 s has been reported (67, p. 55).

Gas Chromatography.   The chromatographic principle is one of separating a mixture into its component substances according to the variation in some suitable physical property of the substances. In gas chromatography two physical properties are involved—solubility and volatility. The chromatographic column consists of a tube packed with a powder of relatively uniform particle size and usually coated with a nonvolatile liquid. Many forms of powder support have been used for oxygen analysis with typical examples being silica gel and molecular sieves. For the liquid a hydrocarbon grease is commonly employed. The column is often maintained at a constant temperature, which is varied according to the gas composition. A carrier gas—usually helium—continually bathes the involatile liquid, and an equilibrium exists between the carrier gas and the amount of gas which has dissolved in the liquid. When oxygen is added to the carrier it is partitioned between the carrier stream and the stationary liquid phase, hence the more complete name of the technique of gas—liquid partition chromatography. At any particular position on the column the equilibrium between the gas in solution in the nonvolatile liquid coating on the particles and as vapor in the interstices of the particles is upset as the vapor fraction moves along the column under the pressure of the carrier gas. So material in solution moves out and vapor passes into solution to form fresh equilibrium positions. The vapor fractions move on again, fresh equilibria are formed, and so the process continues. In this manner a chromatographic zone develops and traverses the column at a speed which depends partly on the pressure of the gas stream and partly on the partitioning between the gaseous and liquid phases. When more than one substance is initially present those with the lowest solubility move faster and the mixture is separated as it passes through the column. If the liquid phase is omitted then some separation may still be achieved due to differential adsorption of the gas on the solid phase bringing about the partitioning.

The block diagram of Fig. 8.16 shows how gas chromatography is used for dissolved oxygen analysis. The stripper is ideally of the type of Fig. 8.9 so that efficient gas exchange is ensured. The drying tube is necessary to prevent problems associated with the slow degassing of water from the column (41). The chromatographic separation step may use more than one column if the mixture is complex, but this should not normally be necessary for oxygen analysis.

A number of different methods of detecting the chromatographic zones as they emerge can be used. A common form of detector is the *ionization detector* which is illustrated in Fig. 8.17. This detector consists of a metal chamber lined with radioactive foil and carrying a central insulated electrode at a high voltage ($\sim 10^3$ V) relative to the chamber wall. When the carrier gas enters the chamber a proportion of it is ionized by radiation from the foil and a steady ionization current flows. There is then an increase in this current when a chromatographic zone enters the detector and this is a measure of the component in the gas stream. The most common detector, however, is the *katharometer*. This is simply a thermal conductivity detector of the type we have already described, and it operates in exactly the same

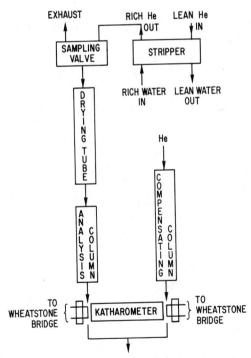

Fig. 8.16 Schematic of oxygen analysis by gas chromatography.

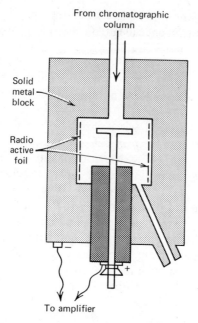

Fig. 8.17   An ionization detector.

manner as when it is used directly for oxygen analysis. In fact, in principle there is no reason why any of the methods of analysis that have been discussed throughout should not be used as a detector for the oxygen after its exit from the column, and gas chromatography can then be regarded as a means of purifying the sample to prevent contamination of the detection device. So, for example, it has been suggested (39, 40) that a Hersch cell [Section 7.4] be used as a detector. Using this electrochemical cell it is claimed that a higher sensitivity can be achieved than with a katharometer (39), and also the cell was found not to be deactivated as rapidly as when it was used in an untreated gas stream (40).

Whatever method of detection is used the output is basically the same, and a typical chromatogram is shown in Fig. 8.18; the detector in this case happens to be a katharometer (41). The calibration is made either by using a standard gas mixture of known oxygen concentration (39), or by coulometric generation to obtain dissolved oxygen to a specific concentration (42). If a gaseous standard is used rather than a dissolved oxygen standard then the calibration and sample peaks may be different in shape, and any comparison must be made on the basis of peak area rather than peak height (41). Various measurements using gas chromatography for dissolved oxygen analysis have been described (71, 72), and details of a portable, field gas chromatograph for gas analysis in lake waters have also recently been given (66).

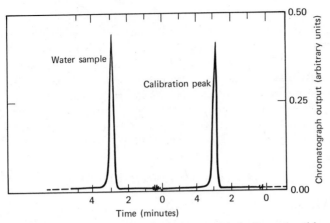

Fig. 8.18 Typical chromatogram for dissolved oxygen analysis. (Reproduced by permission from reference 41. Copyright by the American Chemical Society.)

8.2.3. SUMMARY OF GASEOUS PHASE MEASUREMENTS

The main advantage gained by removing the dissolved oxygen from the sample before monitoring, as we have already indicated, is to remove substances that would otherwise interfere with the method of detection. In addition there are a number of other benefits from gaseous monitoring. Except for manometric and volumetric methods, all the techniques described lend themselves to continuous monitoring and to field operation. Again except for gasometric methods, the various methods can readily be automated and are also capable of responding within very short times of sampling; as already mentioned, the paramagnetic method has a typical response time of a few seconds (67, p. 55). As well as these advantages gas chromatography, specifically, allows simultaneous analysis of other gases besides oxygen and requires particularly low sample volumes. The major general disadvantage of the methods lies in the fact that the apparatus is not always robust and may need considerable protection from shocks and from the environment. But provided a small station is built at the appropriate monitoring site then this should not be too much of a problem. Also, of course, none of the methods are really suitable for *in situ* analysis since it would not be easy to make the equipment completely submersible.

Apart from gasometric methods, which are capable of a truly high precision [$\pm 0.05\%$—(35)], all of the other methods are precise to about $\pm 1\%$ under favorable conditions and when great care is taken (29, 39), that is, not quite as good as the best obtainable by chemical methods. Under less favorable conditions the precision can be rather worse than $\pm 1\%$, and with gas chromatography, for example, it can be as poor as $\pm 10\%$ (58). The sensitivity of the nonchemical methods, on the other hand, is probably higher than that obtained chemically, and with gas chromatography, at least, measurements at the 1 μg liter^{-1} (i.e. 0.001 ppm) level have been reported (43, 44). In view of the fact that the methods described can be readily automated for routine, continuous, on-site analysis, and yet are only slightly less precise than the standard Winkler method, then they offer interesting alternatives or supplements to monitoring with electrochemical techniques. However, they do not have the considerable advantage of MPODs of being able to monitor dissolved oxygen *in situ*.

8.3. RADIOMETRIC METHOD

This method is very similar to the conductometric method using the oxidation of metallic thallium [Section 7.5.1]

$$4Tl + O_2 + 2H_2O \rightarrow 4Tl^+ + 4OH^-$$

except that radioactive metal is used in place of nonactive material. The isotope used is ^{204}Tl which is a β emitter (0.75 MeV) and which has a half life of 3.6 years. The apparatus consists of ^{204}Tl electrodeposited on copper turnings (45) or copper wool (46), and two flow-through-type Geiger— Mueller counters (47). The first counter measures the background β activity and the second the radioactive Tl released by the oxidation reaction; the efficiency of electron capture of the counters is 1–2%. One milligram of dissolved oxygen liberates 25.6 mg of ^{204}Tl and the counting rate is directly proportional to the oxygen concentration in the solution.

The radiometric method is an *in situ* method and, indeed, measurements have been made down to a depth of one mile with no adverse effects (47). Furthermore, equipment has been described (45, 48) which is battery operated and so the technique is suitable for field work, although column preparation is a rather complex procedure. Interferences to the method arise from dissolved salts affecting the Tl(I) solubility (46), from the oxidation being pH dependent, and from other oxidants causing thallium dissolution at low pH (48).

The main disadvantage of the technique is probably the low precision. Because of the randomness of radioactive disintegrations the precision will not be better than $\pm 2\%$ and could be as low as $\pm 5\%$ at 1 ppm (48). Also, in general, the lower limit of detection is about 0.1 ppm with a sensitivity of the same order (29, 48), although results claiming determinations of as little as 1 part per trillion have been reported (45).

Other radiometric methods have been suggested for oxygen analysis, such as the use of dissolved radioactive thallium (49) and of surface layers of radioactive kryptonate of thallium (50), but the method we have just described is the most well tried. A review of applications of radioactivity for trace analysis, including that of dissolved oxygen, has been published (51).

REFERENCES

1. L. W. Winkler, "The determination of dissolved oxygen in water," *Ber. Deut. Chem. Ges.*, **21**, 2843 (1888).

2. H. A. C. Montgomery, N. S. Thom, and A. Cockburn, "Determination of dissolved oxygen by the Winkler method, and the solubility of oxygen in pure water and sea water," *J. Appl. Chem.*, **14**, 280 (1964).

3. *Standard Methods for the Examination of Water and Wastewater*, 13th ed., American Public Health Association Inc., 1971.

4. G. Alsterberg, "Methods for the determination of elementary oxygen dissolved in water in the presence of nitrite," *Biochem. Z.*, **159**, 36 (1925).

5. S. Rideal and G. G. Stewart, "The determination of dissolved oxygen in water in the presence of nitrites and of organic matter," *Analyst*, **26**, 141 (1901).

6. R. Pomeroy and H. D. Kirschman, "Determination of dissolved oxygen: Proposed modifications of the Winkler method," *Anal. Chem.*, **17**, 715 (1945).

7. H. A. C. Montgomery and A. Cockburn, "Errors in sampling for dissolved oxygen," *Analyst*, **89**, 679 (1964).

8. D. E. Carritt and J. H. Carpenter, "Comparison and evaluation of currently employed modifications of the Winkler method for determining dissolved oxygen in sea water," *J. Marine Res.*, **24**, 286 (1966).

9. "The determination of dissolved oxygen," Colorimetric Chemical Analytical Methods, Tintometer Ltd., England.

10. P. A. Hamlin and J. L. Lambert, "Determination of dissolved oxygen using photoreduced leuco phenothiazine dyes," *Anal. Chem.*, **43**, 618 (1971).

11. W. W. Broenkow and J. D. Kline, "Colorimetric determination of dissolved oxygen at low concentrations," *Limnol. Oceanogr.*, **14**, 450 (1969).

12. D. R. Browning, Ed., *Electrometric Methods*, McGraw-Hill, London, 1969.

13. *Manual on Water*, 3rd ed., ASTM Special Technical Publication No. 442, 1969.

14. M. M. Jones and M. W. Mullen, "Some aspects of the Winkler determination of oxygen in water," *Talanta*, **20**, 327 (1973).

15. P. Delahay, *New Instrumental Methods in Electrochemistry*, Interscience, New York, 1954.

16. "Reviews of analytical applications—Water Analysis," *Analytical Chemistry*, 1977, 1975, 1973, 1971, etc.

17. "Annual literature reviews," *Journal of the Water Pollution Control Federation*.

18. O. M. Shtern, "Colorimetric determination of dissolved oxygen at low concentrations," *Energetik*, **12**, 18 (1964).

19. G. P. Sutotskii and Y. M. Gramatchikov, "Rapid determination of small amounts of oxygen," *Teploenergetika*, **13**, 86 (1966).

20.   A. S. Ivanitskaya, "Selection of the optimum method for determining small concentrations of oxygen in water at operating conditions," *Vodopodgot. Vod. Rezhim Khimkontr. Parosilovykh Ustanovkakh,* **1969,** 174.

21.   L. S. Buchoff, N. M. Ingber, and J. H. Brady, "Determination of low concentrations of dissolved oxygen in water," *Anal. Chem.,* **27,** 1401 (1955).

22.   H. Fadrus and J. Maly, "Modified determination of oxygen in water by titration with Fe(II)," *Vom Wasser,* **34,** 132 (1967).

23.   G. S. Sastry, R. E. Hamm, and K. H. Pool, "Spectrophotometric determination of dissolved oxygen in water," *Anal. Chem.,* **41,** 857 (1969).

24.   A. A. Evsina and M. L. Nagibina, "Determination of dissolved oxygen in mine waters," *Zavod. Lab.,* **37,** 784 (1971).

25.   A. Gertner, V. Grdinic, and M. Parag, "Complexometric determination of oxygen in aqueous solutions," *Acta Pharm. Jugoslav.,* **20,** 191 (1970).

26.   I. M. de Graaf Bierbrauwer and H. L. Golterman, "The determination of oxygen in fresh water with trivalent cerium salts," *Chem. Environ. Aquatic Habitat, Proc. IBP Symp.,* **1966,** 158 (1967).

27.   E. J. Green, "Redetermination of the solubility of oxygen in sea water and some thermodynamic implications of the solubility relations," Ph.D. Thesis, Massachusetts Institute of Technology, 1965.

28.   R. Battino and H. L. Clever, "The solubility of gases in liquids," *Chem. Revs.,* **66,** 395 (1966).

29.   K. H. Mancy, Ed., *Instrumental Analysis for Water Pollution Control,* Ann Arbor Science Publishers Inc., Ann Arbor, Mich., 1971.

30.   D. D. Williams and R. R. Miller, "An instrument for on-stream stripping and gas chromatographic determination of dissolved gases in liquids," *Anal. Chem.,* **34,** 657 (1962).

31.   W. J. Moore, *Physical Chemistry,* Longmans, London, 1964.

32.   B. Cantone and S. Gurrieri, "Determination by mass spectrometry of gases dissolved in water," *Bull. Sedute Accud. Gioenia Sci. Nat. Catania,* **72,** 681 (1960).

33.   B. B. Benson and P. D. M. Parker, "Relations among the solubilities of nitrogen, argon and oxygen in distilled water and sea water," *J. Phys. Chem.,* **65,** 1489 (1961).

34.   C. E. Klots and B. B. Benson, "Isotope effects in the solution of oxygen and nitrogen in distilled water," *J. Chem. Phys.,* **38,** 890 (1963).

35.   M. W. Cook and D. N. Hanson, "Accurate measurement of gas solubility," *Rev. Sci. Instr.,* **28,** 370 (1957).

36.   J. J. McKeown, L. C. Brown, and G. W. Gove, "Comparative studies of dissolved oxygen analysis methods," *J. Water Poll. Contr. Fed.,* **39,** 1323 (1967).

37.   J. von Staden and H. Schlageter, "The chlorator phase exchanger, an instrument for continuous measurement and control of oxygen content in strongly contaminated waters," *Oesterr.-Abwasser-Rundsch.,* **10,** 90 (1965).

38.   H. Thiele, "Possibility of continuous measurement of oxygen content in water

and sewage as well as oxygen consumption in systems with high consumption rate," *Industrieabwasser*, **1968**, 22.

39. T. R. Phillips, E. G. Johnson, and H. Woodword, "The use of a Hersch cell as a detector in gas chromatography," *Anal. Chem.*, **36**, 450 (1964).

40. G. E. Hillman and J. Lightwood, "Determination of small amounts of oxygen using a Hersch cell as a gas chromatography detector," *Anal. Chem.*, **38**, 1430 (1966).

41. G. Barbery and Y. Bérubé, "An automated method for analysis of dissolved gases in water at high temperatures and pressures," *Ing. Eng. Chem. Fundam.*, **10**, 632 (1971).

42. H. C. Edgington and R. M. Roberts, "Development of a calibration module for trace oxygen analysers," U.S. Office Saline Water, Res. Devel. Prog. Rept., No. 625, 1970.

43. H. Gaunt and C. Shanks, "Determination of oxygen in boiler feed water," *Chem. Ind. (London)*, **1965**, 328.

44. J. Ropars, "Determination by gas chromatography of traces of oxygen dissolved in steam boiler feed water," *Chim. Anal. (Paris)*, **50**, 641 (1968).

45. A. S. Gillespie and H. G. Richter, "Dissolved oxygen analyser," U.S. Atomic Energy Commission, ORO-592, 1963.

46. J. Rygaert et al., "Radiometric determination of oxygen and chlorine ions dissolved in water," *Ind. Chem. Belge*, **32**, 145 (1967).

47. A. S. Gillespie and K. F. Roberts, "Dissolved oxygen analyser for oceanographic use," U.S. Atomic Energy Commission, ORO-596, 1965.

48. A. S. Gillespie and H. G. Richter, "Radio release instrument for dissolved oxygen analysis," *Anal. Instr.*, **1965**, 127.

49. C. Beaudet and J. Rygaert, "Radiometric determination of ionic chlorine and oxygen in water," *J. Radioanal. Chem.*, **1**, 153 (1968).

50. J. Tolgyessy and S. Varga, "Use of radioactive kryptonate of thallium in the determination of oxygen dissolved in water and other liquids," *Talanta*, **16**, 625 (1969).

51. J. Tolgyessy, "Nuclear analytical methods based on the release and measurement of radioactive reaction products," *Jad. Energ.*, **14**, 169 (1968).

52. B. Ostrom, "Expendable ampoules for oxygen determination," *Mar. Chem.*, **1**, 323 (1973).

53. D. M. Novak and B. E. Conway, "Technique for repetitive gas solubility determinations at various pressures," *Chem. Instr.*, **5**, 79 (1974).

54. J. Miller, "A field method for determining dissolved oxygen in water," *J. Soc. Chem. Ind.*, **33**, 185 (1914).

55. J. Ellis and S. Kenamori, "An evaluation of the Miller method for dissolved oxygen analysis," *Limnol. Oceanogr.*, **18**, 1002 (1973).

56. E. J. Green and D. E. Carritt, "An improved iodine determination flask for whole-bottle titrations," *Analyst*, **9**, 207 (1966).

57. R. Karlsson and L-G. Torstensson, "Coulometric determination of dissolved oxygen in water," *Talanta*, **21**, 957 (1974).

58. R. J. Wilcock and R. Battino, "Solubility of oxygen-nitrogen mixtures in water," *Nature*, **252**, 614 (1974).

59. K. H. Mancy and T. Jaffe, "Analysis of dissolved oxygen in natural and waste waters," U.S. Public Health Service Publ., 999-WP-37, 1966.

60. D. L. King, "Sampling in natural waters," in *Water and Water Pollution Handbook* (L. L. Ciaccio, Ed.), Vol. II, Ch. 10, Marcel Dekker, New York, 1971.

61. ASTM Special Technical Publication No. 148-1, Philadelphia, 1966.

62. U.S. Geological Survey Water Supply Paper No. 1454, U.S. Government Printing Office, Washington, 1960.

63. G. P. Alcock and K. B. Coates, "Indigo carmine for the colorimetric determination of low concentrations of dissolved oxygen in water," *Chem. Ind.*, **10**, 554 (1958).

64. J. Arnott and J. McPheat, "Estimation of dissolved oxygen in de-aerated water," *J. Eng.*, **176**, 103 (1953).

65. P. F. Scholander et al., "Micro-gasometric determination of dissolved oxygen and nitrogen," *Bull. Biol. Woods Hole*, **109**, 328 (1955).

66. D. A. J. Murray and B. van der Veen, "Portable field gas chromatograph for gas analyses in lake water," *Electrochem. Soc. Extended Abstr.*, **75-1**, Abstr. No. 330 (1975).

67. F. Kreuzer, *Oxygen Pressure Recording in Gases, Fluids, and Tissues*, S. Karger, Basel, 1969.

68. D. D. Van Slyke and J. M. Neill, "The determination of gases in blood and other solutions by vacuum extraction and manometric measurement," *J. Biol. Chem.*, **61**, 523 (1924).

69. R. Zander and R. Euler, "Concentration measurement of physically dissolved oxygen by the classical Van Slyke principle," in *Measurement of Oxygen* (H. Degn, I. Balslev, and R. Brook, Eds.), Elsevier Scientific Publishing Company, Amsterdam, 1976.

70. J. McK. Nobbs, "Mass spectrometric determination of oxygen," in *Measurement of Oxygen* (H. Degn, I. Balslev, and R. Brook, Eds.), Elsevier Scientific Publishing Company, Amsterdam, 1976.

71. E. Bjergbakke, "Gas chromatographic measurements of oxygen in aqueous solutions," in *Measurement of Oxygen* (H. Degn, I. Balslev, and R. Brook, Eds.), Elsevier Scientific Publishing Company, Amsterdam, 1976.

72. K. A. Smith, R. J. Dowdell, and K. C. Hall, "Measurement of oxygen in the soil atmosphere and in aqueous solution by gas chromatography," in *Measurement of Oxygen* (H. Degn, I. Balslev, and R. Brook, Eds.), Elsevier Scientific Publishing Company, Amsterdam, 1976.

APPENDIX

A

UNITS

Although there are still a large number of industries which continue to use nonmetric units of measurement, nevertheless the metric system in general and *le Système International* (SI) in particular are being adopted more and more by all areas of industry, trade, and commerce. We have therefore, with only one or two exceptions which we discuss below, used SI units throughout the text.

TABLE A.1   Basic SI Units

Physical quantity	Name of basic SI unit	Symbol for basic SI unit
Length	meter	m
Mass	kilogram	kg
Time	second	s
Electric current	ampere	A
Thermodynamic temperature	kelvin	K
Amount of substance	mole	mol

The basic SI units are listed in Table A.1, and the definitions of these units have been given in a number of places, for example references 1 and 2. Note that the unit kelvin (K) is recommended in place of degree Kelvin ($^\circ$K).

The SI units for certain physical quantities have special names and symbols and some of these are given in Table A.2 together with their definitions. Also given in this table are the SI units for some derived physical quantities which have no special names.

Table A.3 lists prefixes which may be used for decimal fractions or multiples of the basic or derived SI units. It will be seen that on the SI system familiar units such as cm (10^{-2} m), angstrom (10^{-10} m), micron (10^{-6} m), and liter (10^{-3} m^3) are allowed because they are fractions of SI units. However, the names of some of these units, for example, liter or angstrom, are not part of the SI, and it has been recommended that they should be progressively abandoned. Although the basic concepts of the metric system

191

TABLE A.2   Derived SI Units

Physical quantity	Name of SI unit	Symbol	Definition
Energy	joule	J	$kg\ m^2\ s^{-2}$
Force	newton	N	$kg\ m\ s^{-2} = J\ m^{-1}$
Pressure	newton per square meter	1	$N\ m^{-2} = kg\ m^{-1}\ s^{-2} = J\ m^{-3}$
Power	watt	W	$kg\ m^2\ s^{-3} = J\ s^{-1}$
Electric charge	coulomb	C	$A\ s$
Electric potential difference	volt	V	$kg\ m^2\ s^{-3}\ A^{-1} = J\ A^{-1}\ s^{-1}$
Electric resistance	ohm	Ω	$kg\ m^2\ s^{-3}\ A^{-2} = V\ A^{-1}$
Electric capacitance	farad	F	$A^2\ s^4\ kg^{-1}\ m^{-2} = A\ s\ V^{-1}$
Frequency	hertz	Hz	s^{-1}
Density	kilogram per cubic meter		$kg\ m^{-3}$
Viscosity	newton second per square meter		$N\ s\ m^{-2} = kg\ m^{-1}\ s^{-1} = J\ m^{-3}\ s$
Surface tension	newton per meter		$N\ m^{-1} = kg\ s^{-2} = J\ m^{-2}$

[1] The Pascal (Pa) has been suggested.

TABLE A.3   Prefixes for SI Units

Fraction	Prefix	Symbol	Multiple	Prefix	Symbol
10^{-1}	deci	d	10	deka	da
10^{-2}	centi	c	10^2	hecto	h
10^{-3}	milli	m	10^3	kilo	k
10^{-6}	micro	μ	10^6	mega	M
10^{-9}	nano	n	10^9	giga	G
10^{-12}	pico	p	10^{12}	tera	T
10^{-15}	femto	f			
10^{-18}	atto	a			

seem to be slowly being widely accepted, the ingrained habits of language are probably going to be more difficult to overcome (3), and indeed there seems to us to be no virtue in throwing out such well-known terms as angstrom and liter for 0.1 nm and dm^3 just for the sake of conformity. Therefore, in company with crystallographers who seem to be resolutely sticking to angstrom units, we have chosen to retain liters rather than use dm^3.

There are other units which are not part of the SI, but which are very familiar—for example, atmosphere, torr, and calorie. Again it is recommended that these units should eventually no longer be used. But here also the speed with which this demise occurs will depend on how entrenched any given unit is in the common parlance of science and technology and everyday life (3). So, for example, the conversion of calories to joules and vice versa is something one is familiar with from early school science, and thinking in terms of either generally presents no problems. We have used the joule as the unit of energy throughout. On the other hand, although the conversion of degrees Celsius to kelvin is equally simple, because of everyday usage one tends to think in terms of °C (2°C is a cold day and 30°C is a hot one) rather than K. So we have used both °C and K according to which has seemed to be the most appropriate for a given occasion.

The physical quantity which is perhaps the most difficult to think of in terms of SI units is that of pressure. The units of atmosphere (atm), torr (torr), and millimeters of mercury (mm Hg) are very familiar, and most technically trained people have no difficulty in conceiving of pressures of, say, 5 atm or 10^{-2} torr. To think in terms of 506625 N m^{-2} (5 atm) or 1.33 N m^{-2} (10^{-2} torr) is not so easy. We have opted for the easy way out on this one and have retained torr and atm. In the context of pressure measurements it should be noted that the bar—used, for example, in meteorology—is an exact multiple of the SI unit of pressure—1 bar = 10^5 N m^{-2}. Other common units which are now defined exactly in terms of SI units are given in Table A.4.

Units given in figures reproduced from other sources and which do not conform to the SI we have left unchanged.

TABLE A.4   Units Now Exactly Defined in Terms of the SI Units

Physical quantity	Name of unit	Symbol for unit	Definition of unit
Length	inch	in.	2.54×10^{-2} m
Mass	pound (avoirdupois)	lb	0.45359237 kg
Force	kilogram-force	kgf	9.80665 N
Pressure	atmosphere	atm	101325 N m^{-2}
Pressure	torr	torr	$(101325/760)$ N m^{-2}
Pressure	conventional millimeter of mercury	mm Hg	$13.5951 \times 980.665 \times 10^{-2}$ N m^{-2}
Energy	kilowatthour	kWh	3.6×10^6 J
Energy	thermochemical calorie	cal (thermo-chem.)	4.184 J
Common temperature (t)	degree Celsius	°C	$t/°C = T/K - 273.15$
Common temperature (t)	degree Fahrenheit	°F	$t/°F = (9/5)T/K - 459.67$

REFERENCES

1. *Report of the Symbols Committee of the Royal Society*, The Royal Society, London, 1969.
2. M. L. McGlashan, *Physico-chemical Quantities and Units*, Royal Institute of Chemistry, London, 1968.
3. News report, *Scientific American*, **234**, 60A (1976).

B

OXYGEN SOLUBILITY TABLES

Equation (2.15) allows the solubility of oxygen in air-saturated fresh water to be calculated for any temperature and pressure provided that the values of the Bunsen absorption coefficient, α, and the vapor pressure of water, p_1, at the particular temperature are known. Equation (2.10) or (2.11) can be used to obtain α, and p_1 can either be calculated from eq. (2.14) or looked up in tables (1, p. D143). Although the complete calculation is relatively simple, it is nevertheless tedious and so either a nomogram [Fig. 2.5] or a grid is useful. Table B.1 gives oxygen solubilities in mg liter^{-1} for temperature intervals of 1° from 0–50°C and pressure intervals of 5 torr from 600–795 torr. The values have been computed from eq. (2.15) using the coefficients given in Table 2.2 for eq. (2.11) and tabulated values of p_1 (1, p. D143). Interpolation of the data of Carpenter (2) gives solubility values which agree with those in Table B.1 to better than $\pm 0.5\%$ (3).

The dependence of solubility on salt concentration can also be obtained from eq. (2.15) except that now values of α calculated from either eq. (2.23) or (2.24) have to be used. With three variables a nomogram is most convenient—Fig. 2.7—but again a grid can be constructed. Table B.2 lists solubilities for 1° intervals in the range 0–30°C, 10 torr intervals from 600–800 torr, and every 2 g liter^{-1} of chloride from 0–20 g liter^{-1}; if salinity is preferred as a measure of salt concentration then the conversion from g liter^{-1} can be readily made using eq. (2.25). All the intervals are sufficiently small to allow linear interpolation to be used for obtaining intermediate solubility values. The values given in Table B.2 agree to better than $\pm 1\%$ with those of Green and Carritt (4) and of Carpenter (2).

REFERENCES

1. *Handbook of Chemistry and Physics* (51st ed.), The Chemical Rubber Company, Cleveland, 1970.

2. J. H. Carpenter, "New measurements of oxygen solubility in pure and natural water," *Limnol. Oceanogr.*, **11**, 264 (1966).

3. J. M. Hale, personal communication.

4. E. J. Green and D. E. Carritt, "New tables for oxygen saturation of seawater," *J. Mar. Res.*, **25**, 140 (1967).

TABLE B 1

OXYGEN SOLUBILITY IN AIR SATURATED FRESH WATER (MG/L)

TEMP(C)	600	605	610	615	PRESSURE(TORR) 620	625	630	635	640	645
0	11.48	11.58	11.67	11.77	11.87	11.96	12.06	12.16	12.25	12.35
1	11.17	11.26	11.36	11.45	11.54	11.64	11.73	11.83	11.92	12.01
2	10.87	10.96	11.05	11.14	11.23	11.32	11.42	11.51	11.60	11.69
3	10.58	10.67	10.76	10.85	10.94	11.02	11.11	11.20	11.29	11.38
4	10.30	10.39	10.48	10.56	10.65	10.74	10.82	10.91	11.00	11.08
5	10.04	10.12	10.21	10.29	10.37	10.46	10.54	10.63	10.71	10.80
6	9.78	9.86	9.95	10.03	10.11	10.19	10.28	10.36	10.44	10.52
7	9.54	9.62	9.70	9.78	9.86	9.94	10.02	10.10	10.18	10.26
8	9.30	9.38	9.46	9.54	9.61	9.69	9.77	9.85	9.93	10.01
9	9.07	9.15	9.23	9.30	9.38	9.46	9.53	9.61	9.69	9.76
10	8.86	8.93	9.01	9.08	9.16	9.23	9.31	9.38	9.46	9.53
11	8.65	8.72	8.80	8.87	8.94	9.02	9.09	9.16	9.24	9.31
12	8.45	8.52	8.59	8.66	8.74	8.81	8.88	8.95	9.02	9.09
13	8.26	8.33	8.40	8.47	8.54	8.61	8.68	8.75	8.82	8.89
14	8.07	8.14	8.21	8.28	8.35	8.42	8.48	8.55	8.62	8.69
15	7.90	7.96	8.03	8.10	8.16	8.23	8.30	8.37	8.43	8.50
16	7.72	7.79	7.86	7.92	7.99	8.05	8.12	8.19	8.25	8.32
17	7.56	7.63	7.69	7.75	7.82	7.88	7.95	8.01	8.08	8.14
18	7.40	7.47	7.53	7.59	7.66	7.72	7.78	7.85	7.91	7.97
19	7.25	7.31	7.38	7.44	7.50	7.56	7.62	7.69	7.75	7.81
20	7.11	7.17	7.23	7.29	7.35	7.41	7.47	7.53	7.59	7.65
21	6.96	7.02	7.08	7.14	7.20	7.26	7.32	7.38	7.44	7.50
22	6.83	6.89	6.95	7.00	7.06	7.12	7.18	7.24	7.30	7.36
23	6.70	6.76	6.81	6.87	6.93	6.99	7.04	7.10	7.16	7.22
24	6.57	6.63	6.68	6.74	6.80	6.85	6.91	6.97	7.03	7.08
25	6.45	6.50	6.56	6.62	6.67	6.73	6.78	6.84	6.90	6.95
26	6.33	6.38	6.44	6.49	6.55	6.60	6.66	6.71	6.77	6.82
27	6.21	6.27	6.32	6.38	6.43	6.49	6.54	6.59	6.65	6.70
28	6.10	6.16	6.21	6.26	6.32	6.37	6.42	6.48	6.53	6.58
29	5.99	6.05	6.10	6.15	6.21	6.26	6.31	6.36	6.42	6.47
30	5.89	5.94	5.99	6.05	6.10	6.15	6.20	6.25	6.30	6.36
31	5.79	5.84	5.89	5.94	5.99	6.04	6.09	6.15	6.20	6.25
32	5.69	5.74	5.79	5.84	5.89	5.94	5.99	6.04	6.09	6.14
33	5.59	5.64	5.69	5.74	5.79	5.84	5.89	5.94	5.99	6.04
34	5.50	5.54	5.59	5.64	5.69	5.74	5.79	5.84	5.89	5.94
35	5.40	5.45	5.50	5.55	5.60	5.64	5.69	5.74	5.79	5.84
36	5.31	5.36	5.41	5.45	5.50	5.55	5.60	5.65	5.69	5.74
37	5.22	5.27	5.32	5.36	5.41	5.46	5.51	5.55	5.60	5.65
38	5.13	5.18	5.23	5.27	5.32	5.37	5.41	5.46	5.51	5.55
39	5.05	5.09	5.14	5.19	5.23	5.28	5.32	5.37	5.42	5.46
40	4.96	5.01	5.06	5.10	5.15	5.19	5.24	5.28	5.33	5.37
41	4.88	4.92	4.97	5.01	5.06	5.11	5.15	5.20	5.24	5.29
42	4.80	4.84	4.89	4.93	4.98	5.02	5.06	5.11	5.15	5.20
43	4.72	4.76	4.80	4.85	4.89	4.94	4.98	5.02	5.07	5.11
44	4.64	4.68	4.72	4.77	4.81	4.85	4.90	4.94	4.98	5.03
45	4.56	4.60	4.64	4.68	4.73	4.77	4.81	4.86	4.90	4.94
46	4.48	4.52	4.56	4.60	4.65	4.69	4.73	4.77	4.82	4.86
47	4.40	4.44	4.48	4.52	4.57	4.61	4.65	4.69	4.74	4.78
48	4.32	4.36	4.40	4.45	4.49	4.53	4.57	4.61	4.65	4.70
49	4.24	4.28	4.33	4.37	4.41	4.45	4.49	4.53	4.57	4.62
50	4.17	4.21	4.25	4.29	4.33	4.37	4.41	4.45	4.49	4.53

TEMP(C)	PRESSURE(TORR)									
	650	655	660	665	670	675	680	685	690	695
0	12.45	12.54	12.64	12.74	12.83	12.93	13.02	13.12	13.22	13.31
1	12.11	12.20	12.29	12.39	12.48	12.58	12.67	12.76	12.86	12.95
2	11.78	11.87	11.96	12.06	12.15	12.24	12.33	12.42	12.51	12.60
3	11.47	11.56	11.65	11.74	11.83	11.91	12.00	12.09	12.18	12.27
4	11.17	11.26	11.34	11.43	11.52	11.60	11.69	11.78	11.86	11.95
5	10.88	10.97	11.05	11.14	11.22	11.30	11.39	11.47	11.56	11.64
6	10.61	10.69	10.77	10.85	10.94	11.02	11.10	11.18	11.27	11.35
7	10.34	10.42	10.50	10.58	10.66	10.74	10.82	10.90	10.98	11.06
8	10.09	10.16	10.24	10.32	10.40	10.48	10.56	10.64	10.71	10.79
9	9.84	9.92	9.99	10.07	10.15	10.23	10.30	10.38	10.46	10.53
10	9.61	9.68	9.76	9.83	9.91	9.98	10.06	10.13	10.21	10.28
11	9.38	9.46	9.53	9.60	9.68	9.75	9.82	9.89	9.97	10.04
12	9.17	9.24	9.31	9.38	9.45	9.52	9.60	9.67	9.74	9.81
13	8.96	9.03	9.10	9.17	9.24	9.31	9.38	9.45	9.52	9.59
14	8.76	8.83	8.90	8.96	9.03	9.10	9.17	9.24	9.31	9.38
15	8.57	8.63	8.70	8.77	8.84	8.90	8.97	9.04	9.11	9.17
16	8.38	8.45	8.52	8.58	8.65	8.71	8.78	8.84	8.91	8.98
17	8.21	8.27	8.34	8.40	8.46	8.53	8.59	8.66	8.72	8.79
18	8.04	8.10	8.16	8.23	8.29	8.35	8.42	8.48	8.54	8.61
19	7.87	7.93	8.00	8.06	8.12	8.18	8.25	8.31	8.37	8.43
20	7.72	7.78	7.84	7.90	7.96	8.02	8.08	8.14	8.20	8.26
21	7.56	7.62	7.68	7.74	7.80	7.86	7.92	7.98	8.04	8.10
22	7.42	7.48	7.53	7.59	7.65	7.71	7.77	7.83	7.89	7.95
23	7.28	7.33	7.39	7.45	7.51	7.56	7.62	7.68	7.74	7.80
24	7.14	7.20	7.25	7.31	7.37	7.42	7.48	7.54	7.59	7.65
25	7.01	7.06	7.12	7.18	7.23	7.29	7.34	7.40	7.45	7.51
26	6.88	6.93	6.99	7.05	7.10	7.16	7.21	7.27	7.32	7.38
27	6.76	6.81	6.86	6.92	6.97	7.03	7.08	7.14	7.19	7.24
28	6.64	6.69	6.74	6.80	6.85	6.90	6.96	7.01	7.06	7.12
29	6.52	6.57	6.63	6.68	6.73	6.78	6.84	6.89	6.94	6.99
30	6.41	6.46	6.51	6.56	6.62	6.67	6.72	6.77	6.82	6.87
31	6.30	6.35	6.40	6.45	6.50	6.55	6.61	6.66	6.71	6.76
32	6.19	6.24	6.29	6.34	6.39	6.44	6.49	6.54	6.59	6.65
33	6.09	6.14	6.19	6.24	6.29	6.34	6.39	6.44	6.49	6.54
34	5.99	6.04	6.08	6.13	6.18	6.23	6.28	6.33	6.38	6.43
35	5.89	5.94	5.98	6.03	6.08	6.13	6.18	6.23	6.27	6.32
36	5.79	5.84	5.89	5.93	5.98	6.03	6.08	6.12	6.17	6.22
37	5.69	5.74	5.79	5.84	5.88	5.93	5.98	6.03	6.07	6.12
38	5.60	5.65	5.69	5.74	5.79	5.83	5.88	5.93	5.97	6.02
39	5.51	5.56	5.60	5.65	5.69	5.74	5.79	5.83	5.88	5.92
40	5.42	5.47	5.51	5.56	5.60	5.65	5.69	5.74	5.78	5.83
41	5.33	5.38	5.42	5.47	5.51	5.56	5.60	5.65	5.69	5.74
42	5.24	5.29	5.33	5.38	5.42	5.47	5.51	5.55	5.60	5.64
43	5.16	5.20	5.24	5.29	5.33	5.38	5.42	5.46	5.51	5.55
44	5.07	5.11	5.16	5.20	5.25	5.29	5.33	5.38	5.42	5.46
45	4.99	5.03	5.07	5.12	5.16	5.20	5.25	5.29	5.33	5.37
46	4.90	4.95	4.99	5.03	5.07	5.12	5.16	5.20	5.24	5.29
47	4.82	4.86	4.90	4.95	4.99	5.03	5.07	5.12	5.16	5.20
48	4.74	4.78	4.82	4.86	4.91	4.95	4.99	5.03	5.07	5.11
49	4.66	4.70	4.74	4.78	4.82	4.86	4.91	4.95	4.99	5.03
50	4.58	4.62	4.66	4.70	4.74	4.78	4.82	4.86	4.90	4.95

TEMP(C)

	700	705	710	715	PRESSURE(TORR) 720	725	730	735	740	745
0	13.41	13.51	13.60	13.70	13.80	13.89	13.99	14.08	14.18	14.28
1	13.05	13.14	13.23	13.33	13.42	13.51	13.61	13.70	13.80	13.89
2	12.70	12.79	12.88	12.97	13.06	13.15	13.24	13.34	13.43	13.52
3	12.36	12.45	12.54	12.63	12.72	12.80	12.89	12.98	13.07	13.16
4	12.04	12.12	12.21	12.30	12.38	12.47	12.56	12.64	12.73	12.82
5	11.73	11.81	11.90	11.98	12.07	12.15	12.23	12.32	12.40	12.49
6	11.43	11.51	11.60	11.68	11.76	11.84	11.93	12.01	12.09	12.17
7	11.15	11.23	11.31	11.39	11.47	11.55	11.63	11.71	11.79	11.87
8	10.87	10.95	11.03	11.11	11.19	11.26	11.34	11.42	11.50	11.58
9	10.61	10.69	10.76	10.84	10.92	10.99	11.07	11.15	11.22	11.30
10	10.36	10.43	10.51	10.58	10.66	10.73	10.81	10.88	10.96	11.03
11	10.11	10.19	10.26	10.33	10.41	10.48	10.55	10.63	10.70	10.77
12	9.88	9.95	10.03	10.10	10.17	10.24	10.31	10.38	10.46	10.53
13	9.66	9.73	9.80	9.87	9.94	10.01	10.08	10.15	10.22	10.29
14	9.45	9.51	9.58	9.65	9.72	9.79	9.86	9.93	9.99	10.06
15	9.24	9.31	9.37	9.44	9.51	9.58	9.64	9.71	9.78	9.84
16	9.04	9.11	9.17	9.24	9.31	9.37	9.44	9.50	9.57	9.64
17	8.85	8.92	8.98	9.05	9.11	9.18	9.24	9.30	9.37	9.43
18	8.67	8.73	8.80	8.86	8.92	8.99	9.05	9.11	9.18	9.24
19	8.49	8.56	8.62	8.68	8.74	8.80	8.87	8.93	8.99	9.05
20	8.33	8.39	8.45	8.51	8.57	8.63	8.69	8.75	8.81	8.87
21	8.16	8.22	8.28	8.34	8.40	8.46	8.52	8.58	8.64	8.70
22	8.01	8.06	8.12	8.18	8.24	8.30	8.36	8.42	8.48	8.53
23	7.85	7.91	7.97	8.03	8.09	8.14	8.20	8.26	8.32	8.37
24	7.71	7.76	7.82	7.88	7.94	7.99	8.05	8.11	8.16	8.22
25	7.57	7.62	7.68	7.73	7.79	7.85	7.90	7.96	8.01	8.07
26	7.43	7.49	7.54	7.60	7.65	7.71	7.76	7.82	7.87	7.93
27	7.30	7.35	7.41	7.46	7.52	7.57	7.62	7.68	7.73	7.79
28	7.17	7.22	7.28	7.33	7.38	7.44	7.49	7.54	7.60	7.65
29	7.05	7.10	7.15	7.20	7.26	7.31	7.36	7.41	7.47	7.52
30	6.93	6.98	7.03	7.08	7.13	7.19	7.24	7.29	7.34	7.39
31	6.81	6.86	6.91	6.96	7.01	7.06	7.12	7.17	7.22	7.27
32	6.70	6.75	6.80	6.85	6.90	6.95	7.00	7.05	7.10	7.15
33	6.58	6.63	6.68	6.73	6.78	6.83	6.88	6.93	6.98	7.03
34	6.48	6.53	6.57	6.62	6.67	6.72	6.77	6.82	6.87	6.92
35	6.37	6.42	6.47	6.52	6.56	6.61	6.66	6.71	6.76	6.81
36	6.27	6.32	6.36	6.41	6.46	6.51	6.55	6.60	6.65	6.70
37	6.17	6.21	6.26	6.31	6.36	6.40	6.45	6.50	6.54	6.59
38	6.07	6.11	6.16	6.21	6.25	6.30	6.35	6.39	6.44	6.49
39	5.97	6.02	6.06	6.11	6.15	6.20	6.25	6.29	6.34	6.39
40	5.88	5.92	5.97	6.01	6.06	6.10	6.15	6.19	6.24	6.29
41	5.78	5.83	5.87	5.92	5.96	6.01	6.05	6.10	6.14	6.19
42	5.69	5.73	5.78	5.82	5.87	5.91	5.96	6.00	6.04	6.09
43	5.60	5.64	5.68	5.73	5.77	5.82	5.86	5.91	5.95	5.99
44	5.51	5.55	5.59	5.64	5.68	5.72	5.77	5.81	5.86	5.90
45	5.42	5.46	5.50	5.55	5.59	5.63	5.68	5.72	5.76	5.81
46	5.33	5.37	5.42	5.46	5.50	5.54	5.59	5.63	5.67	5.71
47	5.24	5.28	5.33	5.37	5.41	5.45	5.50	5.54	5.58	5.62
48	5.16	5.20	5.24	5.28	5.32	5.37	5.41	5.45	5.49	5.53
49	5.07	5.11	5.15	5.20	5.24	5.28	5.32	5.36	5.40	5.44
50	4.99	5.03	5.07	5.11	5.15	5.19	5.23	5.27	5.31	5.36

TEMP(C)	750	755	760	765	PRESSURE(TORR) 770	775	780	785	790	795
0	14.37	14.47	14.57	14.66	14.76	14.86	14.95	15.05	15.15	15.24
1	13.98	14.08	14.17	14.27	14.36	14.45	14.55	14.64	14.73	14.83
2	13.61	13.70	13.79	13.88	13.97	14.07	14.16	14.25	14.34	14.43
3	13.25	13.34	13.43	13.52	13.61	13.69	13.78	13.87	13.96	14.05
4	12.90	12.99	13.08	13.16	13.25	13.34	13.42	13.51	13.60	13.68
5	12.57	12.66	12.74	12.83	12.91	13.00	13.08	13.16	13.25	13.33
6	12.25	12.34	12.42	12.50	12.58	12.67	12.75	12.83	12.91	13.00
7	11.95	12.03	12.11	12.19	12.27	12.35	12.43	12.51	12.59	12.67
8	11.66	11.74	11.81	11.89	11.97	12.05	12.13	12.21	12.29	12.36
9	11.38	11.45	11.53	11.61	11.68	11.76	11.84	11.91	11.99	12.07
10	11.11	11.18	11.26	11.33	11.41	11.48	11.56	11.63	11.71	11.78
11	10.85	10.92	10.99	11.07	11.14	11.21	11.29	11.36	11.43	11.51
12	10.60	10.67	10.74	10.81	10.89	10.96	11.03	11.10	11.17	11.24
13	10.36	10.43	10.50	10.57	10.64	10.71	10.78	10.85	10.92	10.99
14	10.13	10.20	10.27	10.34	10.41	10.48	10.54	10.61	10.68	10.75
15	9.91	9.98	10.05	10.11	10.18	10.25	10.32	10.38	10.45	10.52
16	9.70	9.77	9.83	9.90	9.96	10.03	10.10	10.16	10.23	10.29
17	9.50	9.56	9.63	9.69	9.76	9.82	9.89	9.95	10.01	10.08
18	9.30	9.37	9.43	9.49	9.56	9.62	9.68	9.75	9.81	9.87
19	9.12	9.18	9.24	9.30	9.36	9.43	9.49	9.55	9.61	9.67
20	8.93	9.00	9.06	9.12	9.18	9.24	9.30	9.36	9.42	9.48
21	8.76	8.82	8.88	8.94	9.00	9.06	9.12	9.18	9.24	9.30
22	8.59	8.65	8.71	8.77	8.83	8.89	8.95	9.01	9.06	9.12
23	8.43	8.49	8.55	8.61	8.66	8.72	8.78	8.84	8.90	8.95
24	8.28	8.33	8.39	8.45	8.50	8.56	8.62	8.67	8.73	8.79
25	8.13	8.18	8.24	8.29	8.35	8.41	8.46	8.52	8.57	8.63
26	7.98	8.04	8.09	8.15	8.20	8.26	8.31	8.37	8.42	8.48
27	7.84	7.89	7.95	8.00	8.06	8.11	8.17	8.22	8.27	8.33
28	7.70	7.76	7.81	7.86	7.92	7.97	8.02	8.08	8.13	8.18
29	7.57	7.63	7.68	7.73	7.78	7.84	7.89	7.94	7.99	8.05
30	7.44	7.50	7.55	7.60	7.65	7.70	7.76	7.81	7.86	7.91
31	7.32	7.37	7.42	7.47	7.52	7.58	7.63	7.68	7.73	7.78
32	7.20	7.25	7.30	7.35	7.40	7.45	7.50	7.55	7.60	7.65
33	7.08	7.13	7.18	7.23	7.28	7.33	7.38	7.43	7.48	7.53
34	6.97	7.02	7.07	7.11	7.16	7.21	7.26	7.31	7.36	7.41
35	6.86	6.90	6.95	7.00	7.05	7.10	7.15	7.19	7.24	7.29
36	6.75	6.79	6.84	6.89	6.94	6.98	7.03	7.08	7.13	7.18
37	6.64	6.69	6.73	6.78	6.83	6.88	6.92	6.97	7.02	7.06
38	6.53	6.58	6.63	6.67	6.72	6.77	6.81	6.86	6.91	6.95
39	6.43	6.48	6.52	6.57	6.62	6.66	6.71	6.75	6.80	6.85
40	6.33	6.38	6.42	6.47	6.51	6.56	6.60	6.65	6.70	6.74
41	6.23	6.28	6.32	6.37	6.41	6.46	6.50	6.55	6.59	6.64
42	6.13	6.18	6.22	6.27	6.31	6.36	6.40	6.45	6.49	6.53
43	6.04	6.08	6.13	6.17	6.21	6.26	6.30	6.35	6.39	6.43
44	5.94	5.99	6.03	6.07	6.12	6.16	6.20	6.25	6.29	6.33
45	5.85	5.89	5.94	5.98	6.02	6.06	6.11	6.15	6.19	6.24
46	5.76	5.80	5.84	5.88	5.93	5.97	6.01	6.06	6.10	6.14
47	5.67	5.71	5.75	5.79	5.83	5.88	5.92	5.96	6.00	6.05
48	5.57	5.62	5.66	5.70	5.74	5.78	5.83	5.87	5.91	5.95
49	5.49	5.53	5.57	5.61	5.65	5.69	5.73	5.78	5.82	5.86
50	5.40	5.44	5.48	5.52	5.56	5.60	5.64	5.68	5.72	5.77

199

TABLE B2

OXYGEN SOLUBILITY IN AIR SATURATED SALINE WATER (MG/L)

PRESSURE(TORR)= 600

TEMP(C)   CHLORIDE(G/L)

	0	2	4	6	8	10	12	14	16	18	20
0	11.48	11.22	10.96	10.71	10.45	10.19	9.93	9.67	9.41	9.15	8.90
1	11.17	10.92	10.67	10.42	10.17	9.92	9.67	9.42	9.17	8.92	8.67
2	10.87	10.63	10.38	10.14	9.90	9.66	9.41	9.17	8.93	8.69	8.45
3	10.58	10.35	10.11	9.88	9.64	9.41	9.17	8.94	8.70	8.47	8.24
4	10.30	10.08	9.85	9.62	9.40	9.17	8.94	8.72	8.49	8.26	8.04
5	10.04	9.82	9.60	9.38	9.16	8.94	8.72	8.50	8.28	8.06	7.84
6	9.78	9.57	9.36	9.14	8.93	8.72	8.51	8.30	8.08	7.87	7.66
7	9.54	9.33	9.13	8.92	8.72	8.51	8.31	8.10	7.90	7.69	7.49
8	9.30	9.10	8.90	8.71	8.51	8.31	8.11	7.91	7.71	7.52	7.32
9	9.07	8.88	8.69	8.50	8.31	8.12	7.92	7.73	7.54	7.35	7.16
10	8.86	8.67	8.49	8.30	8.12	7.93	7.75	7.56	7.37	7.19	7.00
11	8.65	8.47	8.29	8.11	7.93	7.75	7.57	7.39	7.22	7.04	6.86
12	8.45	8.28	8.10	7.93	7.76	7.58	7.41	7.24	7.06	6.89	6.72
13	8.26	8.09	7.92	7.75	7.59	7.42	7.25	7.08	6.91	6.75	6.58
14	8.07	7.91	7.75	7.59	7.42	7.26	7.10	6.94	6.77	6.61	6.45
15	7.90	7.74	7.58	7.42	7.27	7.11	6.95	6.79	6.64	6.48	6.32
16	7.72	7.57	7.42	7.27	7.11	6.96	6.81	6.66	6.51	6.35	6.20
17	7.56	7.41	7.27	7.12	6.97	6.82	6.67	6.53	6.38	6.23	6.08
18	7.40	7.26	7.12	6.97	6.83	6.69	6.54	6.40	6.25	6.11	5.97
19	7.25	7.11	6.97	6.83	6.69	6.55	6.41	6.27	6.13	5.99	5.86
20	7.11	6.97	6.83	6.70	6.56	6.43	6.29	6.15	6.02	5.88	5.75
21	6.96	6.83	6.70	6.57	6.43	6.30	6.17	6.04	5.90	5.77	5.64
22	6.83	6.70	6.57	6.44	6.31	6.18	6.05	5.92	5.79	5.66	5.53
23	6.70	6.57	6.44	6.32	6.19	6.06	5.94	5.81	5.68	5.56	5.43
24	6.57	6.45	6.32	6.20	6.07	5.95	5.82	5.70	5.58	5.45	5.33
25	6.45	6.33	6.20	6.08	5.96	5.84	5.71	5.59	5.47	5.35	5.23
26	6.33	6.21	6.09	5.97	5.85	5.73	5.61	5.49	5.36	5.24	5.12
27	6.21	6.10	5.98	5.86	5.74	5.62	5.50	5.38	5.26	5.14	5.02
28	6.10	5.98	5.87	5.75	5.63	5.51	5.39	5.27	5.16	5.04	4.92
29	5.99	5.88	5.76	5.64	5.52	5.41	5.29	5.17	5.05	4.93	4.82
30	5.89	5.77	5.65	5.54	5.42	5.30	5.18	5.07	4.95	4.83	4.71

200

PRESSURE(TORR)= 610

TEMP(C)   CHLORIDE(G/L)

	0	2	4	6	8	10	12	14	16	18	20
0	11.67	11.41	11.15	10.89	10.62	10.36	10.10	9.83	9.57	9.31	9.05
1	11.36	11.10	10.85	10.59	10.34	10.08	9.83	9.58	9.32	9.07	8.81
2	11.05	10.80	10.56	10.31	10.07	9.82	9.57	9.33	9.08	8.83	8.59
3	10.76	10.52	10.28	10.04	9.80	9.57	9.33	9.09	8.85	8.61	8.37
4	10.48	10.25	10.01	9.78	9.55	9.32	9.09	8.86	8.63	8.40	8.17
5	10.21	9.98	9.76	9.54	9.31	9.09	8.87	8.64	8.42	8.20	7.98
6	9.95	9.73	9.51	9.30	9.08	8.87	8.65	8.44	8.22	8.01	7.79
7	9.70	9.49	9.28	9.07	8.86	8.65	8.45	8.24	8.03	7.82	7.61
8	9.46	9.26	9.05	8.85	8.65	8.45	8.25	8.05	7.84	7.64	7.44
9	9.23	9.03	8.84	8.64	8.45	8.25	8.06	7.86	7.67	7.47	7.28
10	9.01	8.82	8.63	8.44	8.25	8.06	7.88	7.69	7.50	7.31	7.12
11	8.80	8.61	8.43	8.25	8.07	7.88	7.70	7.52	7.34	7.16	6.97
12	8.59	8.42	8.24	8.06	7.89	7.71	7.53	7.36	7.18	7.01	6.83
13	8.40	8.23	8.06	7.89	7.71	7.54	7.37	7.20	7.03	6.86	6.69
14	8.21	8.04	7.88	7.71	7.55	7.38	7.22	7.05	6.89	6.72	6.56
15	8.03	7.87	7.71	7.55	7.39	7.23	7.07	6.91	6.75	6.59	6.43
16	7.86	7.70	7.55	7.39	7.24	7.08	6.93	6.77	6.62	6.46	6.31
17	7.69	7.54	7.39	7.24	7.09	6.94	6.79	6.64	6.49	6.34	6.19
18	7.53	7.38	7.24	7.09	6.95	6.80	6.65	6.51	6.36	6.22	6.07
19	7.38	7.23	7.09	6.95	6.81	6.67	6.52	6.38	6.24	6.10	5.96
20	7.23	7.09	6.95	6.81	6.67	6.54	6.40	6.26	6.12	5.98	5.84
21	7.08	6.95	6.81	6.68	6.54	6.41	6.28	6.14	6.01	5.87	5.74
22	6.95	6.81	6.68	6.55	6.42	6.29	6.16	6.02	5.89	5.76	5.63
23	6.81	6.68	6.56	6.43	6.30	6.17	6.04	5.91	5.78	5.65	5.52
24	6.68	6.56	6.43	6.30	6.18	6.05	5.93	5.80	5.67	5.55	5.42
25	6.56	6.44	6.31	6.19	6.06	5.94	5.81	5.69	5.56	5.44	5.32
26	6.44	6.32	6.19	6.07	5.95	5.83	5.70	5.58	5.46	5.34	5.21
27	6.32	6.20	6.08	5.96	5.84	5.72	5.59	5.47	5.35	5.23	5.11
28	6.21	6.09	5.97	5.85	5.73	5.61	5.49	5.37	5.25	5.13	5.01
29	6.10	5.98	5.86	5.74	5.62	5.50	5.38	5.26	5.14	5.02	4.90
30	5.99	5.87	5.75	5.63	5.51	5.39	5.28	5.16	5.04	4.92	4.80

PRESSURE(TORR)= 620

TEMP(C)   CHLORIDE(G/L)

	0	2	4	6	8	10	12	14	16	18	20
0	11.87	11.60	11.33	11.07	10.80	10.53	10.26	10.00	9.73	9.46	9.20
1	11.54	11.29	11.03	10.77	10.51	10.25	9.99	9.73	9.47	9.22	8.96
2	11.23	10.98	10.73	10.48	10.23	9.98	9.73	9.48	9.23	8.98	8.73
3	10.94	10.69	10.45	10.21	9.97	9.72	9.48	9.24	9.00	8.76	8.51
4	10.65	10.41	10.18	9.95	9.71	9.48	9.24	9.01	8.77	8.54	8.31
5	10.37	10.15	9.92	9.69	9.47	9.24	9.01	8.79	8.56	8.33	8.11
6	10.11	9.89	9.67	9.45	9.23	9.01	8.80	8.58	8.36	8.14	7.92
7	9.86	9.65	9.43	9.22	9.01	8.80	8.59	8.37	8.16	7.95	7.74
8	9.61	9.41	9.20	9.00	8.79	8.59	8.38	8.18	7.98	7.77	7.57
9	9.38	9.18	8.98	8.79	8.59	8.39	8.19	7.99	7.80	7.60	7.40
10	9.16	8.97	8.77	8.58	8.39	8.20	8.01	7.82	7.62	7.43	7.24
11	8.94	8.76	8.57	8.39	8.20	8.02	7.83	7.65	7.46	7.27	7.09
12	8.74	8.56	8.38	8.20	8.02	7.84	7.66	7.48	7.30	7.12	6.94
13	8.54	8.36	8.19	8.02	7.84	7.67	7.50	7.32	7.15	6.98	6.80
14	8.35	8.18	8.01	7.84	7.68	7.51	7.34	7.17	7.00	6.84	6.67
15	8.16	8.00	7.84	7.68	7.51	7.35	7.19	7.03	6.86	6.70	6.54
16	7.99	7.83	7.67	7.52	7.36	7.20	7.04	6.88	6.73	6.57	6.41
17	7.82	7.67	7.51	7.36	7.21	7.05	6.90	6.75	6.60	6.44	6.29
18	7.66	7.51	7.36	7.21	7.06	6.91	6.77	6.62	6.47	6.32	6.17
19	7.50	7.36	7.21	7.07	6.92	6.78	6.63	6.49	6.34	6.20	6.06
20	7.35	7.21	7.07	6.93	6.79	6.65	6.51	6.37	6.22	6.08	5.94
21	7.20	7.07	6.93	6.79	6.66	6.52	6.38	6.24	6.11	5.97	5.83
22	7.06	6.93	6.80	6.66	6.53	6.39	6.26	6.13	5.99	5.86	5.72
23	6.93	6.80	6.67	6.54	6.40	6.27	6.14	6.01	5.88	5.75	5.62
24	6.80	6.67	6.54	6.41	6.28	6.15	6.03	5.90	5.77	5.64	5.51
25	6.67	6.55	6.42	6.29	6.17	6.04	5.91	5.79	5.66	5.53	5.41
26	6.55	6.42	6.30	6.18	6.05	5.93	5.80	5.68	5.55	5.43	5.30
27	6.43	6.31	6.18	6.06	5.94	5.81	5.69	5.57	5.44	5.32	5.20
28	6.32	6.19	6.07	5.95	5.83	5.70	5.58	5.46	5.34	5.21	5.09
29	6.21	6.08	5.96	5.84	5.72	5.60	5.47	5.35	5.23	5.11	4.99
30	6.10	5.98	5.85	5.73	5.61	5.49	5.37	5.24	5.12	5.00	4.88

201

PRESSURE(TORR)= 630

TEMP(C)   CHLORIDE(G/L)

	0	2	4	6	8	10	12	14	16	18	20
0	12.06	11.79	11.52	11.25	10.97	10.70	10.43	10.16	9.89	9.62	9.34
1	11.73	11.47	11.21	10.94	10.68	10.42	10.15	9.89	9.63	9.37	9.10
2	11.42	11.16	10.91	10.65	10.40	10.14	9.89	9.64	9.38	9.13	8.87
3	11.11	10.87	10.62	10.37	10.13	9.88	9.64	9.39	9.14	8.90	8.65
4	10.82	10.58	10.35	10.11	9.87	9.63	9.39	9.16	8.92	8.68	8.44
5	10.54	10.31	10.08	9.85	9.62	9.39	9.16	8.93	8.70	8.47	8.24
6	10.28	10.05	9.83	9.61	9.38	9.16	8.94	8.72	8.49	8.27	8.05
7	10.02	9.80	9.59	9.37	9.16	8.94	8.73	8.51	8.30	8.08	7.86
8	9.77	9.56	9.36	9.15	8.94	8.73	8.52	8.31	8.11	7.90	7.69
9	9.53	9.33	9.13	8.93	8.73	8.53	8.33	8.12	7.92	7.72	7.52
10	9.31	9.11	8.92	8.72	8.53	8.33	8.14	7.94	7.75	7.55	7.36
11	9.09	8.90	8.71	8.52	8.34	8.15	7.96	7.77	7.58	7.39	7.21
12	8.88	8.70	8.51	8.33	8.15	7.97	7.79	7.60	7.42	7.24	7.06
13	8.68	8.50	8.33	8.15	7.97	7.80	7.62	7.44	7.27	7.09	6.91
14	8.48	8.31	8.14	7.97	7.80	7.63	7.46	7.29	7.12	6.95	6.78
15	8.30	8.13	7.97	7.80	7.64	7.47	7.31	7.14	6.98	6.81	6.65
16	8.12	7.96	7.80	7.64	7.48	7.32	7.16	7.00	6.84	6.68	6.52
17	7.95	7.79	7.64	7.48	7.33	7.17	7.02	6.86	6.70	6.55	6.39
18	7.78	7.63	7.48	7.33	7.18	7.03	6.88	6.73	6.58	6.42	6.27
19	7.62	7.48	7.33	7.18	7.04	6.89	6.74	6.60	6.45	6.30	6.16
20	7.47	7.33	7.19	7.04	6.90	6.76	6.61	6.47	6.33	6.18	6.04
21	7.32	7.18	7.04	6.91	6.77	6.63	6.49	6.35	6.21	6.07	5.93
22	7.18	7.05	6.91	6.77	6.64	6.50	6.36	6.23	6.09	5.96	5.82
23	7.04	6.91	6.78	6.64	6.51	6.38	6.24	6.11	5.98	5.84	5.71
24	6.91	6.78	6.65	6.52	6.39	6.26	6.13	6.00	5.87	5.73	5.60
25	6.78	6.65	6.53	6.40	6.27	6.14	6.01	5.88	5.75	5.63	5.50
26	6.66	6.53	6.41	6.28	6.15	6.03	5.90	5.77	5.64	5.52	5.39
27	6.54	6.41	6.29	6.16	6.04	5.91	5.79	5.66	5.54	5.41	5.28
28	6.42	6.30	6.17	6.05	5.93	5.80	5.68	5.55	5.43	5.30	5.18
29	6.31	6.19	6.06	5.94	5.81	5.69	5.57	5.44	5.32	5.19	5.07
30	6.20	6.08	5.95	5.83	5.71	5.58	5.46	5.33	5.21	5.09	4.96

PRESSURE(TORR)= 640

TEMP(C)   CHLORIDE(G/L)

	0	2	4	6	8	10	12	14	16	18	20
0	12.25	11.98	11.70	11.43	11.15	10.87	10.60	10.32	10.05	9.77	9.49
1	11.92	11.65	11.39	11.12	10.85	10.58	10.32	10.05	9.78	9.52	9.25
2	11.60	11.34	11.08	10.82	10.57	10.31	10.05	9.79	9.53	9.27	9.01
3	11.29	11.04	10.79	10.54	10.29	10.04	9.79	9.54	9.29	9.04	8.79
4	11.00	10.75	10.51	10.27	10.03	9.79	9.54	9.30	9.06	8.82	8.58
5	10.71	10.48	10.24	10.01	9.78	9.54	9.31	9.07	8.84	8.61	8.37
6	10.44	10.21	9.99	9.76	9.54	9.31	9.08	8.86	8.63	8.40	8.18
7	10.18	9.96	9.74	9.52	9.30	9.09	8.87	8.65	8.43	8.21	7.99
8	9.93	9.72	9.51	9.29	9.08	8.87	8.66	8.45	8.24	8.02	7.81
9	9.69	9.48	9.28	9.07	8.87	8.66	8.46	8.26	8.05	7.85	7.64
10	9.46	9.26	9.06	8.86	8.67	8.47	8.27	8.07	7.87	7.68	7.48
11	9.24	9.04	8.85	8.66	8.47	8.28	8.09	7.90	7.70	7.51	7.32
12	9.02	8.84	8.65	8.47	8.28	8.10	7.91	7.73	7.54	7.36	7.17
13	8.82	8.64	8.46	8.28	8.10	7.92	7.74	7.56	7.38	7.21	7.03
14	8.62	8.45	8.27	8.10	7.93	7.75	7.58	7.41	7.23	7.06	6.89
15	8.43	8.27	8.10	7.93	7.76	7.59	7.42	7.26	7.09	6.92	6.75
16	8.25	8.09	7.93	7.76	7.60	7.44	7.27	7.11	6.95	6.79	6.62
17	8.08	7.92	7.76	7.60	7.45	7.29	7.13	6.97	6.81	6.66	6.50
18	7.91	7.76	7.60	7.45	7.30	7.14	6.99	6.84	6.68	6.53	6.38
19	7.75	7.60	7.45	7.30	7.15	7.00	6.85	6.70	6.55	6.41	6.26
20	7.59	7.45	7.30	7.16	7.01	6.87	6.72	6.58	6.43	6.29	6.14
21	7.44	7.30	7.16	7.02	6.88	6.74	6.59	6.45	6.31	6.17	6.03
22	7.30	7.16	7.02	6.88	6.75	6.61	6.47	6.33	6.19	6.05	5.92
23	7.16	7.02	6.89	6.75	6.62	6.48	6.35	6.21	6.08	5.94	5.81
24	7.03	6.89	6.76	6.63	6.49	6.36	6.23	6.09	5.96	5.83	5.70
25	6.90	6.76	6.63	6.50	6.37	6.24	6.11	5.98	5.85	5.72	5.59
26	6.77	6.64	6.51	6.38	6.25	6.12	6.00	5.87	5.74	5.61	5.48
27	6.65	6.52	6.39	6.27	6.14	6.01	5.88	5.76	5.63	5.50	5.37
28	6.53	6.40	6.28	6.15	6.02	5.90	5.77	5.64	5.52	5.39	5.26
29	6.42	6.29	6.16	6.04	5.91	5.79	5.66	5.53	5.41	5.28	5.16
30	6.30	6.18	6.05	5.93	5.80	5.67	5.55	5.42	5.30	5.17	5.05

PRESSURE(TORR)= 650

TEMP(C)   CHLORIDE(G/L)

	0	2	4	6	8	10	12	14	16	18	20
0	12.45	12.17	11.89	11.61	11.32	11.04	10.76	10.48	10.20	9.92	9.64
1	12.11	11.84	11.56	11.29	11.02	10.75	10.48	10.21	9.94	9.67	9.39
2	11.78	11.52	11.26	10.99	10.73	10.47	10.21	9.94	9.68	9.42	9.16
3	11.47	11.22	10.96	10.71	10.45	10.20	9.95	9.69	9.44	9.18	8.93
4	11.17	10.92	10.68	10.43	10.19	9.94	9.70	9.45	9.20	8.96	8.71
5	10.88	10.64	10.41	10.17	9.93	9.69	9.46	9.22	8.98	8.74	8.50
6	10.61	10.38	10.15	9.92	9.69	9.46	9.23	9.00	8.77	8.54	8.31
7	10.34	10.12	9.90	9.67	9.45	9.23	9.01	8.78	8.56	8.34	8.12
8	10.09	9.87	9.66	9.44	9.23	9.01	8.80	8.58	8.37	8.15	7.94
9	9.84	9.63	9.43	9.22	9.01	8.80	8.59	8.39	8.18	7.97	7.76
10	9.61	9.41	9.20	9.00	8.80	8.60	8.40	8.20	8.00	7.80	7.60
11	9.38	9.19	8.99	8.80	8.60	8.41	8.22	8.02	7.83	7.63	7.44
12	9.17	8.98	8.79	8.60	8.41	8.23	8.04	7.85	7.66	7.47	7.28
13	8.96	8.78	8.59	8.41	8.23	8.05	7.87	7.68	7.50	7.32	7.14
14	8.76	8.58	8.41	8.23	8.05	7.88	7.70	7.53	7.35	7.17	7.00
15	8.57	8.40	8.23	8.06	7.88	7.71	7.54	7.37	7.20	7.03	6.86
16	8.38	8.22	8.05	7.89	7.72	7.56	7.39	7.23	7.06	6.89	6.73
17	8.21	8.05	7.89	7.73	7.56	7.40	7.24	7.08	6.92	6.76	6.60
18	8.04	7.88	7.72	7.57	7.41	7.26	7.10	6.95	6.79	6.63	6.48
19	7.87	7.72	7.57	7.42	7.27	7.11	6.96	6.81	6.66	6.51	6.36
20	7.72	7.57	7.42	7.27	7.12	6.98	6.83	6.68	6.53	6.39	6.24
21	7.56	7.42	7.28	7.13	6.99	6.84	6.70	6.56	6.41	6.27	6.12
22	7.42	7.28	7.14	6.99	6.85	6.71	6.57	6.43	6.29	6.15	6.01
23	7.28	7.14	7.00	6.86	6.72	6.59	6.45	6.31	6.17	6.04	5.90
24	7.14	7.00	6.87	6.73	6.60	6.46	6.33	6.19	6.06	5.92	5.79
25	7.01	6.87	6.74	6.61	6.48	6.34	6.21	6.08	5.94	5.81	5.68
26	6.88	6.75	6.62	6.49	6.36	6.22	6.09	5.96	5.83	5.70	5.57
27	6.76	6.63	6.50	6.37	6.24	6.11	5.98	5.85	5.72	5.59	5.46
28	6.64	6.51	6.38	6.25	6.12	5.99	5.86	5.74	5.61	5.48	5.35
29	6.52	6.39	6.26	6.14	6.01	5.88	5.75	5.62	5.50	5.37	5.24
30	6.41	6.28	6.15	6.02	5.90	5.77	5.64	5.51	5.38	5.26	5.13

PRESSURE(TORR)= 660

TEMP(C)   CHLORIDE(G/L)

	0	2	4	6	8	10	12	14	16	18	20
0	12.64	12.35	12.07	11.78	11.50	11.22	10.93	10.65	10.36	10.08	9.79
1	12.29	12.02	11.74	11.47	11.19	10.92	10.64	10.37	10.09	9.82	9.54
2	11.96	11.70	11.43	11.16	10.90	10.63	10.36	10.10	9.83	9.56	9.30
3	11.65	11.39	11.13	10.87	10.62	10.36	10.10	9.84	9.58	9.33	9.07
4	11.34	11.09	10.84	10.59	10.34	10.10	9.85	9.60	9.35	9.10	8.85
5	11.05	10.81	10.57	10.33	10.09	9.84	9.60	9.36	9.12	8.88	8.64
6	10.77	10.54	10.30	10.07	9.84	9.60	9.37	9.14	8.90	8.67	8.44
7	10.50	10.28	10.05	9.82	9.60	9.37	9.15	8.92	8.70	8.47	8.24
8	10.24	10.02	9.81	9.59	9.37	9.15	8.93	8.71	8.50	8.28	8.06
9	9.99	9.78	9.57	9.36	9.15	8.94	8.73	8.52	8.31	8.09	7.88
10	9.76	9.55	9.35	9.14	8.94	8.74	8.53	8.33	8.12	7.92	7.72
11	9.53	9.33	9.13	8.94	8.74	8.54	8.34	8.15	7.95	7.75	7.55
12	9.31	9.12	8.93	8.74	8.55	8.35	8.16	7.97	7.78	7.59	7.40
13	9.10	8.91	8.73	8.54	8.36	8.17	7.99	7.80	7.62	7.43	7.25
14	8.90	8.72	8.54	8.36	8.18	8.00	7.82	7.64	7.46	7.29	7.11
15	8.70	8.53	8.36	8.18	8.01	7.84	7.66	7.49	7.31	7.14	6.97
16	8.52	8.35	8.18	8.01	7.84	7.67	7.51	7.34	7.17	7.00	6.83
17	8.34	8.17	8.01	7.85	7.68	7.52	7.36	7.19	7.03	6.87	6.71
18	8.16	8.00	7.85	7.69	7.53	7.37	7.21	7.05	6.90	6.74	6.58
19	8.00	7.84	7.69	7.54	7.38	7.23	7.07	6.92	6.77	6.61	6.46
20	7.84	7.69	7.54	7.39	7.24	7.09	6.94	6.79	6.64	6.49	6.34
21	7.68	7.54	7.39	7.24	7.10	6.95	6.81	6.66	6.51	6.37	6.22
22	7.53	7.39	7.25	7.11	6.96	6.82	6.68	6.53	6.39	6.25	6.11
23	7.39	7.25	7.11	6.97	6.83	6.69	6.55	6.41	6.27	6.13	5.99
24	7.25	7.12	6.98	6.84	6.70	6.57	6.43	6.29	6.16	6.02	5.88
25	7.12	6.98	6.85	6.71	6.58	6.44	6.31	6.17	6.04	5.90	5.77
26	6.99	6.86	6.72	6.59	6.46	6.32	6.19	6.06	5.92	5.79	5.66
27	6.86	6.73	6.60	6.47	6.34	6.21	6.07	5.94	5.81	5.68	5.55
28	6.74	6.61	6.48	6.35	6.22	6.09	5.96	5.83	5.70	5.57	5.44
29	6.63	6.50	6.37	6.24	6.11	5.98	5.84	5.71	5.58	5.45	5.32
30	6.51	6.38	6.25	6.12	5.99	5.86	5.73	5.60	5.47	5.34	5.21

PRESSURE(TORR)= 670

TEMP(C) CHLORIDE(G/L)

	0	2	4	6	8	10	12	14	16	18	20
0	12.83	12.54	12.25	11.96	11.68	11.39	11.10	10.81	10.52	10.23	9.94
1	12.48	12.20	11.92	11.64	11.36	11.08	10.80	10.52	10.24	9.97	9.69
2	12.15	11.88	11.61	11.34	11.06	10.79	10.52	10.25	9.98	9.71	9.44
3	11.83	11.56	11.30	11.04	10.78	10.52	10.25	9.99	9.73	9.47	9.21
4	11.52	11.26	11.01	10.76	10.50	10.25	10.00	9.74	9.49	9.24	8.98
5	11.22	10.97	10.73	10.48	10.24	9.99	9.75	9.50	9.26	9.01	8.77
6	10.94	10.70	10.46	10.22	9.99	9.75	9.51	9.28	9.04	8.80	8.56
7	10.66	10.43	10.20	9.97	9.75	9.52	9.29	9.06	8.83	8.60	8.37
8	10.40	10.18	9.96	9.73	9.51	9.29	9.07	8.85	8.63	8.40	8.18
9	10.15	9.93	9.72	9.51	9.29	9.08	8.86	8.65	8.43	8.22	8.00
10	9.91	9.70	9.49	9.28	9.08	8.87	8.66	8.46	8.25	8.04	7.83
11	9.68	9.47	9.27	9.07	8.87	8.67	8.47	8.27	8.07	7.87	7.67
12	9.45	9.26	9.06	8.87	8.68	8.48	8.29	8.09	7.90	7.71	7.51
13	9.24	9.05	8.86	8.68	8.49	8.30	8.11	7.92	7.74	7.55	7.36
14	9.03	8.85	8.67	8.49	8.31	8.12	7.94	7.76	7.58	7.40	7.22
15	8.84	8.66	8.48	8.31	8.13	7.96	7.78	7.60	7.43	7.25	7.08
16	8.65	8.48	8.31	8.14	7.96	7.79	7.62	7.45	7.28	7.11	6.94
17	8.46	8.30	8.13	7.97	7.80	7.64	7.47	7.31	7.14	6.97	6.81
18	8.29	8.13	7.97	7.81	7.65	7.49	7.32	7.16	7.00	6.84	6.68
19	8.12	7.96	7.81	7.65	7.50	7.34	7.18	7.03	6.87	6.71	6.56
20	7.96	7.81	7.65	7.50	7.35	7.20	7.05	6.89	6.74	6.59	6.44
21	7.80	7.65	7.51	7.36	7.21	7.06	6.91	6.76	6.61	6.47	6.32
22	7.65	7.51	7.36	7.22	7.07	6.93	6.78	6.64	6.49	6.35	6.20
23	7.51	7.36	7.22	7.08	6.94	6.80	6.65	6.51	6.37	6.23	6.09
24	7.37	7.23	7.09	6.95	6.81	6.67	6.53	6.39	6.25	6.11	5.97
25	7.23	7.09	6.96	6.82	6.68	6.55	6.41	6.27	6.13	6.00	5.86
26	7.10	6.96	6.83	6.69	6.56	6.42	6.29	6.15	6.02	5.88	5.75
27	6.97	6.84	6.71	6.57	6.44	6.30	6.17	6.04	5.90	5.77	5.64
28	6.85	6.72	6.58	6.45	6.32	6.19	6.05	5.92	5.79	5.66	5.52
29	6.73	6.60	6.47	6.33	6.20	6.07	5.94	5.81	5.67	5.54	5.41
30	6.62	6.48	6.35	6.22	6.09	5.95	5.82	5.69	5.56	5.43	5.29

PRESSURE(TORR)= 680

TEMP(C) CHLORIDE(G/L)

	0	2	4	6	8	10	12	14	16	18	20
0	13.02	12.73	12.44	12.14	11.85	11.56	11.26	10.97	10.68	10.38	10.09
1	12.67	12.39	12.10	11.82	11.53	11.25	10.97	10.68	10.40	10.12	9.83
2	12.33	12.06	11.78	11.51	11.23	10.96	10.68	10.41	10.13	9.86	9.58
3	12.00	11.74	11.47	11.21	10.94	10.67	10.41	10.14	9.88	9.61	9.34
4	11.69	11.43	11.18	10.92	10.66	10.40	10.15	9.89	9.63	9.38	9.12
5	11.39	11.14	10.89	10.64	10.39	10.15	9.90	9.65	9.40	9.15	8.90
6	11.10	10.86	10.62	10.38	10.14	9.90	9.66	9.42	9.18	8.93	8.69
7	10.82	10.59	10.36	10.13	9.89	9.66	9.43	9.19	8.96	8.73	8.50
8	10.56	10.33	10.11	9.88	9.66	9.43	9.21	8.98	8.76	8.53	8.31
9	10.30	10.08	9.87	9.65	9.43	9.21	9.00	8.78	8.56	8.34	8.13
10	10.06	9.85	9.64	9.43	9.22	9.00	8.79	8.58	8.37	8.16	7.95
11	9.82	9.62	9.41	9.21	9.01	8.80	8.60	8.40	8.19	7.99	7.79
12	9.60	9.40	9.20	9.01	8.81	8.61	8.41	8.22	8.02	7.82	7.63
13	9.38	9.19	9.00	8.81	8.62	8.43	8.24	8.05	7.85	7.66	7.47
14	9.17	8.99	8.80	8.62	8.43	8.25	8.06	7.88	7.69	7.51	7.33
15	8.97	8.79	8.61	8.43	8.26	8.08	7.90	7.72	7.54	7.36	7.18
16	8.78	8.61	8.43	8.26	8.09	7.91	7.74	7.57	7.39	7.22	7.05
17	8.59	8.43	8.26	8.09	7.92	7.75	7.59	7.42	7.25	7.08	6.91
18	8.42	8.25	8.09	7.93	7.76	7.60	7.44	7.27	7.11	6.95	6.78
19	8.25	8.09	7.93	7.77	7.61	7.45	7.29	7.13	6.98	6.82	6.66
20	8.08	7.93	7.77	7.62	7.46	7.31	7.15	7.00	6.84	6.69	6.54
21	7.92	7.77	7.62	7.47	7.32	7.17	7.02	6.87	6.72	6.57	6.41
22	7.77	7.62	7.48	7.33	7.18	7.03	6.89	6.74	6.59	6.44	6.30
23	7.62	7.48	7.33	7.19	7.05	6.90	6.76	6.61	6.47	6.32	6.18
24	7.48	7.34	7.20	7.06	6.91	6.77	6.63	6.49	6.35	6.21	6.07
25	7.34	7.20	7.06	6.93	6.79	6.65	6.51	6.37	6.23	6.09	5.95
26	7.21	7.07	6.94	6.80	6.66	6.52	6.39	6.25	6.11	5.97	5.84
27	7.08	6.95	6.81	6.67	6.54	6.40	6.27	6.13	5.99	5.86	5.72
28	6.96	6.82	6.69	6.55	6.42	6.28	6.15	6.01	5.88	5.74	5.61
29	6.84	6.70	6.57	6.43	6.30	6.16	6.03	5.90	5.76	5.63	5.49
30	6.72	6.58	6.45	6.32	6.18	6.05	5.91	5.78	5.65	5.51	5.38

PRESSURE(TORR)= 690

TEMP(C)   CHLORIDE(G/L)

	0	2	4	6	8	10	12	14	16	18	20
0	13.22	12.92	12.62	12.32	12.03	11.73	11.43	11.13	10.84	10.54	10.24
1	12.86	12.57	12.28	11.99	11.71	11.42	11.13	10.84	10.55	10.26	9.98
2	12.51	12.23	11.96	11.68	11.40	11.12	10.84	10.56	10.28	10.00	9.72
3	12.18	11.91	11.64	11.37	11.10	10.83	10.56	10.29	10.02	9.75	9.48
4	11.86	11.60	11.34	11.08	10.82	10.56	10.30	10.04	9.78	9.51	9.25
5	11.56	11.31	11.05	10.80	10.55	10.30	10.04	9.79	9.54	9.29	9.03
6	11.27	11.02	10.78	10.53	10.29	10.04	9.80	9.56	9.31	9.07	8.82
7	10.98	10.75	10.51	10.28	10.04	9.80	9.57	9.33	9.09	8.86	8.62
8	10.71	10.49	10.26	10.03	9.80	9.57	9.34	9.12	8.89	8.66	8.43
9	10.46	10.23	10.01	9.79	9.57	9.35	9.13	8.91	8.69	8.47	8.25
10	10.21	9.99	9.78	9.57	9.35	9.14	8.93	8.71	8.50	8.28	8.07
11	9.97	9.76	9.56	9.35	9.14	8.94	8.73	8.52	8.32	8.11	7.90
12	9.74	9.54	9.34	9.14	8.94	8.74	8.54	8.34	8.14	7.94	7.74
13	9.52	9.33	9.13	8.94	8.75	8.55	8.36	8.17	7.97	7.78	7.59
14	9.31	9.12	8.93	8.75	8.56	8.37	8.18	8.00	7.81	7.62	7.44
15	9.11	8.92	8.74	8.56	8.38	8.20	8.02	7.84	7.65	7.47	7.29
16	8.91	8.73	8.56	8.38	8.21	8.03	7.86	7.68	7.50	7.33	7.15
17	8.72	8.55	8.38	8.21	8.04	7.87	7.70	7.53	7.36	7.19	7.02
18	8.54	8.38	8.21	8.05	7.88	7.71	7.55	7.38	7.22	7.05	6.89
19	8.37	8.21	8.05	7.89	7.73	7.56	7.40	7.24	7.08	6.92	6.76
20	8.20	8.05	7.89	7.73	7.58	7.42	7.26	7.10	6.95	6.79	6.63
21	8.04	7.89	7.74	7.58	7.43	7.28	7.12	6.97	6.82	6.66	6.51
22	7.89	7.74	7.59	7.44	7.29	7.14	6.99	6.84	6.69	6.54	6.39
23	7.74	7.59	7.45	7.30	7.15	7.01	6.86	6.71	6.57	6.42	6.27
24	7.59	7.45	7.31	7.16	7.02	6.88	6.73	6.59	6.44	6.30	6.16
25	7.45	7.31	7.17	7.03	6.89	6.75	6.61	6.47	6.32	6.18	6.04
26	7.32	7.18	7.04	6.90	6.76	6.62	6.48	6.34	6.20	6.07	5.93
27	7.19	7.05	6.91	6.78	6.64	6.50	6.36	6.22	6.09	5.95	5.81
28	7.06	6.93	6.79	6.65	6.52	6.38	6.24	6.11	5.97	5.83	5.69
29	6.94	6.81	6.67	6.53	6.40	6.26	6.12	5.99	5.85	5.71	5.58
30	6.82	6.69	6.55	6.41	6.28	6.14	6.01	5.87	5.73	5.60	5.46

PRESSURE(TORR)= 700

TEMP(C)   CHLORIDE(G/L)

	0	2	4	6	8	10	12	14	16	18	20
0	13.41	13.11	12.81	12.50	12.20	11.90	11.60	11.30	10.99	10.69	10.39
1	13.05	12.75	12.46	12.17	11.88	11.58	11.29	11.00	10.71	10.41	10.12
2	12.70	12.41	12.13	11.85	11.56	11.28	11.00	10.72	10.43	10.15	9.87
3	12.36	12.09	11.81	11.54	11.26	10.99	10.72	10.44	10.17	9.90	9.62
4	12.04	11.77	11.51	11.24	10.98	10.71	10.45	10.18	9.92	9.65	9.39
5	11.73	11.47	11.21	10.96	10.70	10.45	10.19	9.93	9.68	9.42	9.17
6	11.43	11.18	10.93	10.69	10.44	10.19	9.94	9.70	9.45	9.20	8.95
7	11.15	10.91	10.67	10.43	10.19	9.95	9.71	9.47	9.23	8.99	8.75
8	10.87	10.64	10.41	10.18	9.94	9.71	9.48	9.25	9.02	8.79	8.55
9	10.61	10.38	10.16	9.94	9.71	9.49	9.26	9.04	8.82	8.59	8.37
10	10.36	10.14	9.92	9.71	9.49	9.27	9.06	8.84	8.62	8.41	8.19
11	10.11	9.91	9.70	9.49	9.28	9.07	8.86	8.65	8.44	8.23	8.02
12	9.88	9.68	9.48	9.27	9.07	8.87	8.67	8.46	8.26	8.06	7.85
13	9.66	9.46	9.27	9.07	8.87	8.68	8.48	8.29	8.09	7.89	7.70
14	9.45	9.26	9.07	8.88	8.69	8.50	8.31	8.12	7.93	7.74	7.55
15	9.24	9.06	8.87	8.69	8.50	8.32	8.14	7.95	7.77	7.58	7.40
16	9.04	8.86	8.69	8.51	8.33	8.15	7.97	7.79	7.61	7.44	7.26
17	8.85	8.68	8.51	8.33	8.16	7.99	7.81	7.64	7.47	7.29	7.12
18	8.67	8.50	8.33	8.17	8.00	7.83	7.66	7.49	7.32	7.16	6.99
19	8.49	8.33	8.17	8.00	7.84	7.68	7.51	7.35	7.19	7.02	6.86
20	8.33	8.17	8.01	7.85	7.69	7.53	7.37	7.21	7.05	6.89	6.73
21	8.16	8.01	7.85	7.70	7.54	7.39	7.23	7.07	6.92	6.76	6.61
22	8.01	7.85	7.70	7.55	7.40	7.25	7.09	6.94	6.79	6.64	6.49
23	7.85	7.71	7.56	7.41	7.26	7.11	6.96	6.81	6.67	6.52	6.37
24	7.71	7.56	7.42	7.27	7.12	6.98	6.85	6.69	6.54	6.40	6.25
25	7.57	7.42	7.28	7.14	6.99	6.85	6.71	6.56	6.42	6.28	6.13
26	7.43	7.29	7.15	7.01	6.86	6.72	6.58	6.44	6.30	6.16	6.01
27	7.30	7.16	7.02	6.88	6.74	6.60	6.46	6.32	6.18	6.04	5.90
28	7.17	7.03	6.89	6.75	6.61	6.48	6.34	6.20	6.06	5.92	5.78
29	7.05	6.91	6.77	6.63	6.49	6.35	6.22	6.08	5.94	5.80	5.66
30	6.93	6.79	6.65	6.51	6.37	6.23	6.10	5.96	5.82	5.68	5.54

PRESSURE(TORR)= 710

TEMP(C)   CHLORIDE(G/L)

	0	2	4	6	8	10	12	14	16	18	20
0	13.60	13.30	12.99	12.68	12.38	12.07	11.77	11.46	11.15	10.85	10.54
1	13.23	12.94	12.64	12.34	12.05	11.75	11.45	11.16	10.86	10.56	10.27
2	12.88	12.59	12.30	12.02	11.73	11.44	11.16	10.87	10.58	10.30	10.01
3	12.54	12.26	11.98	11.70	11.43	11.15	10.87	10.59	10.32	10.04	9.76
4	12.21	11.94	11.67	11.40	11.14	10.87	10.60	10.33	10.06	9.79	9.52
5	11.90	11.64	11.38	11.12	10.86	10.60	10.34	10.08	9.82	9.56	9.30
6	11.60	11.34	11.09	10.84	10.59	10.34	10.09	9.84	9.58	9.33	9.08
7	11.31	11.06	10.82	10.58	10.33	10.09	9.85	9.60	9.36	9.12	8.88
8	11.03	10.79	10.56	10.32	10.09	9.85	9.62	9.38	9.15	8.91	8.68
9	10.76	10.53	10.31	10.08	9.85	9.63	9.40	9.17	8.94	8.72	8.49
10	10.51	10.29	10.07	9.85	9.63	9.41	9.19	8.97	8.75	8.53	8.31
11	10.26	10.05	9.84	9.62	9.41	9.20	8.99	8.77	8.56	8.35	8.13
12	10.03	9.82	9.61	9.41	9.20	9.00	8.79	8.59	8.38	8.17	7.97
13	9.80	9.60	9.40	9.20	9.00	8.80	8.61	8.41	8.21	8.01	7.81
14	9.58	9.39	9.20	9.00	8.81	8.62	8.43	8.23	8.04	7.85	7.65
15	9.37	9.19	9.00	8.81	8.63	8.44	8.25	8.07	7.88	7.69	7.51
16	9.17	8.99	8.81	8.63	8.45	8.27	8.09	7.91	7.73	7.54	7.36
17	8.98	8.81	8.63	8.45	8.28	8.10	7.93	7.75	7.58	7.40	7.22
18	8.80	8.63	8.46	8.28	8.11	7.94	7.77	7.60	7.43	7.26	7.09
19	8.62	8.45	8.29	8.12	7.95	7.79	7.62	7.46	7.29	7.12	6.96
20	8.45	8.29	8.12	7.96	7.80	7.64	7.48	7.32	7.15	6.99	6.83
21	8.28	8.12	7.97	7.81	7.65	7.49	7.34	7.18	7.02	6.86	6.71
22	8.12	7.97	7.81	7.66	7.51	7.35	7.20	7.04	6.89	6.74	6.58
23	7.97	7.82	7.67	7.52	7.37	7.22	7.06	6.91	6.76	6.61	6.46
24	7.82	7.67	7.53	7.38	7.23	7.08	6.93	6.79	6.64	6.49	6.34
25	7.68	7.53	7.39	7.24	7.10	6.95	6.81	6.66	6.51	6.37	6.22
26	7.54	7.40	7.25	7.11	6.97	6.82	6.68	6.54	6.39	6.25	6.10
27	7.41	7.26	7.12	6.98	6.84	6.70	6.55	6.41	6.27	6.13	5.99
28	7.28	7.14	7.00	6.85	6.71	6.57	6.43	6.29	6.15	6.01	5.87
29	7.15	7.01	6.87	6.73	6.59	6.45	6.31	6.17	6.03	5.89	5.75
30	7.03	6.89	6.75	6.61	6.47	6.33	6.19	6.05	5.91	5.77	5.63

PRESSURE(TORR)= 720

TEMP(C)   CHLORIDE(G/L)

	0	2	4	6	8	10	12	14	16	18	20
0	13.80	13.49	13.17	12.86	12.55	12.24	11.93	11.62	11.31	11.00	10.69
1	13.42	13.12	12.82	12.52	12.22	11.92	11.62	11.32	11.02	10.71	10.41
2	13.06	12.77	12.48	12.19	11.90	11.61	11.31	11.02	10.73	10.44	10.15
3	12.72	12.43	12.15	11.87	11.59	11.31	11.03	10.74	10.46	10.18	9.90
4	12.38	12.11	11.84	11.57	11.29	11.02	10.75	10.48	10.20	9.93	9.66
5	12.07	11.80	11.54	11.27	11.01	10.75	10.48	10.22	9.96	9.69	9.43
6	11.76	11.51	11.25	11.00	10.74	10.49	10.23	9.98	9.72	9.47	9.21
7	11.47	11.22	10.97	10.73	10.48	10.23	9.99	9.74	9.49	9.25	9.00
8	11.19	10.95	10.71	10.47	10.23	9.99	9.76	9.52	9.28	9.04	8.80
9	10.92	10.69	10.45	10.22	9.99	9.76	9.53	9.30	9.07	8.84	8.61
10	10.66	10.43	10.21	9.99	9.76	9.54	9.32	9.10	8.87	8.65	8.43
11	10.41	10.19	9.98	9.76	9.55	9.33	9.11	8.90	8.68	8.47	8.25
12	10.17	9.96	9.75	9.54	9.33	9.13	8.92	8.71	8.50	8.29	8.08
13	9.94	9.74	9.54	9.33	9.13	8.93	8.73	8.53	8.32	8.12	7.92
14	9.72	9.52	9.33	9.13	8.94	8.74	8.55	8.35	8.16	7.96	7.76
15	9.51	9.32	9.13	8.94	8.75	8.56	8.37	8.18	7.99	7.80	7.61
16	9.31	9.12	8.94	8.75	8.57	8.39	8.20	8.02	7.84	7.65	7.47
17	9.11	8.93	8.75	8.58	8.40	8.22	8.04	7.86	7.68	7.51	7.33
18	8.92	8.75	8.58	8.40	8.23	8.06	7.88	7.71	7.54	7.37	7.19
19	8.74	8.57	8.41	8.24	8.07	7.90	7.73	7.56	7.40	7.23	7.06
20	8.57	8.41	8.24	8.08	7.91	7.75	7.59	7.42	7.26	7.09	6.93
21	8.40	8.24	8.08	7.92	7.76	7.60	7.44	7.28	7.12	6.96	6.80
22	8.24	8.08	7.93	7.77	7.62	7.46	7.30	7.15	6.99	6.83	6.68
23	8.09	7.93	7.78	7.63	7.47	7.32	7.17	7.01	6.86	6.71	6.56
24	7.94	7.79	7.64	7.48	7.33	7.18	7.03	6.88	6.73	6.58	6.43
25	7.79	7.64	7.50	7.35	7.20	7.05	6.90	6.76	6.61	6.46	6.31
26	7.65	7.50	7.36	7.21	7.07	6.92	6.78	6.63	6.48	6.34	6.19
27	7.52	7.37	7.23	7.08	6.94	6.79	6.65	6.51	6.36	6.22	6.07
28	7.38	7.24	7.10	6.95	6.81	6.67	6.53	6.38	6.24	6.10	5.95
29	7.26	7.11	6.97	6.83	6.69	6.54	6.40	6.26	6.12	5.97	5.83
30	7.13	6.99	6.85	6.71	6.57	6.42	6.28	6.14	5.99	5.85	5.71

PRESSURE(TORR)= 730

TEMP(C)   CHLORIDE(G/L)

	0	2	4	6	8	10	12	14	16	18	20
0	13.99	13.67	13.36	13.04	12.73	12.41	12.10	11.78	11.47	11.15	10.84
1	13.61	13.30	13.00	12.69	12.39	12.08	11.78	11.47	11.17	10.86	10.56
2	13.24	12.95	12.65	12.36	12.06	11.77	11.47	11.18	10.88	10.59	10.29
3	12.89	12.61	12.32	12.04	11.75	11.47	11.18	10.89	10.61	10.32	10.04
4	12.56	12.28	12.00	11.73	11.45	11.18	10.90	10.62	10.35	10.07	9.79
5	12.23	11.97	11.70	11.43	11.17	10.90	10.63	10.36	10.10	9.83	9.56
6	11.93	11.67	11.41	11.15	10.89	10.63	10.37	10.12	9.86	9.60	9.34
7	11.63	11.38	11.13	10.88	10.63	10.38	10.13	9.88	9.63	9.38	9.13
8	11.34	11.10	10.86	10.62	10.38	10.13	9.89	9.65	9.41	9.17	8.92
9	11.07	10.84	10.60	10.37	10.13	9.90	9.67	9.43	9.20	8.96	8.73
10	10.81	10.58	10.35	10.13	9.90	9.68	9.45	9.22	9.00	8.77	8.55
11	10.55	10.34	10.12	9.90	9.68	9.46	9.24	9.02	8.80	8.59	8.37
12	10.31	10.10	9.89	9.68	9.47	9.25	9.04	8.83	8.62	8.41	8.20
13	10.08	9.88	9.67	9.47	9.26	9.06	8.85	8.65	8.44	8.24	8.03
14	9.86	9.66	9.46	9.26	9.06	8.87	8.67	8.47	8.27	8.07	7.87
15	9.64	9.45	9.26	9.07	8.87	8.68	8.49	8.30	8.11	7.91	7.72
16	9.44	9.25	9.06	8.88	8.69	8.51	8.32	8.13	7.95	7.76	7.57
17	9.24	9.06	8.88	8.70	8.52	8.34	8.16	7.97	7.79	7.61	7.43
18	9.05	8.87	8.70	8.52	8.35	8.17	8.00	7.82	7.65	7.47	7.29
19	8.87	8.70	8.53	8.35	8.18	8.01	7.84	7.67	7.50	7.33	7.16
20	8.69	8.52	8.36	8.19	8.03	7.86	7.69	7.53	7.36	7.19	7.03
21	8.52	8.36	8.20	8.04	7.87	7.71	7.55	7.39	7.22	7.06	6.90
22	8.36	8.20	8.04	7.88	7.72	7.57	7.41	7.25	7.09	6.93	6.77
23	8.20	8.05	7.89	7.74	7.58	7.43	7.27	7.11	6.96	6.80	6.65
24	8.05	7.90	7.74	7.59	7.44	7.29	7.14	6.98	6.83	6.68	6.53
25	7.90	7.75	7.60	7.45	7.30	7.15	7.00	6.85	6.70	6.55	6.40
26	7.76	7.61	7.47	7.32	7.17	7.02	6.87	6.73	6.58	6.43	6.28
27	7.62	7.48	7.33	7.18	7.04	6.89	6.75	6.60	6.45	6.31	6.16
28	7.49	7.35	7.20	7.06	6.91	6.76	6.62	6.47	6.33	6.18	6.04
29	7.36	7.22	7.07	6.93	6.78	6.64	6.49	6.35	6.21	6.06	5.92
30	7.24	7.09	6.95	6.80	6.66	6.51	6.37	6.23	6.08	5.94	5.79

PRESSURE(TORR)= 740

TEMP(C)   CHLORIDE(G/L)

	0	2	4	6	8	10	12	14	16	18	20
0	14.18	13.86	13.54	13.22	12.90	12.58	12.27	11.95	11.63	11.31	10.99
1	13.80	13.49	13.18	12.87	12.56	12.25	11.94	11.63	11.32	11.01	10.70
2	13.43	13.13	12.83	12.53	12.23	11.93	11.63	11.33	11.03	10.73	10.43
3	13.07	12.78	12.49	12.20	11.91	11.62	11.33	11.04	10.76	10.47	10.18
4	12.73	12.45	12.17	11.89	11.61	11.33	11.05	10.77	10.49	10.21	9.93
5	12.40	12.13	11.86	11.59	11.32	11.05	10.78	10.51	10.24	9.97	9.69
6	12.09	11.83	11.57	11.30	11.04	10.78	10.52	10.26	9.99	9.73	9.47
7	11.79	11.54	11.28	11.03	10.77	10.52	10.27	10.01	9.76	9.51	9.25
8	11.50	11.25	11.01	10.76	10.52	10.27	10.03	9.78	9.54	9.29	9.05
9	11.22	10.99	10.75	10.51	10.27	10.04	9.80	9.56	9.33	9.09	8.85
10	10.96	10.73	10.50	10.27	10.04	9.81	9.58	9.35	9.12	8.89	8.66
11	10.70	10.48	10.26	10.04	9.81	9.59	9.37	9.15	8.93	8.71	8.48
12	10.46	10.24	10.03	9.81	9.60	9.38	9.17	8.95	8.74	8.52	8.31
13	10.22	10.01	9.81	9.60	9.39	9.18	8.97	8.77	8.56	8.35	8.14
14	9.99	9.79	9.59	9.39	9.19	8.99	8.79	8.59	8.39	8.18	7.98
15	9.78	9.58	9.39	9.19	9.00	8.80	8.61	8.41	8.22	8.02	7.83
16	9.57	9.38	9.19	9.00	8.81	8.62	8.44	8.25	8.06	7.87	7.68
17	9.37	9.19	9.00	8.82	8.64	8.45	8.27	8.09	7.90	7.72	7.54
18	9.18	9.00	8.82	8.64	8.46	8.29	8.11	7.93	7.75	7.57	7.40
19	8.99	8.82	8.64	8.47	8.30	8.13	7.95	7.78	7.61	7.43	7.26
20	8.81	8.64	8.48	8.31	8.14	7.97	7.80	7.63	7.46	7.30	7.13
21	8.64	8.48	8.31	8.15	7.98	7.82	7.65	7.49	7.33	7.16	7.00
22	8.48	8.32	8.15	7.99	7.83	7.67	7.51	7.35	7.19	7.03	6.87
23	8.32	8.16	8.00	7.84	7.69	7.53	7.37	7.22	7.06	6.90	6.74
24	8.16	8.01	7.85	7.70	7.55	7.39	7.24	7.08	6.93	6.77	6.62
25	8.01	7.86	7.71	7.56	7.41	7.25	7.10	6.95	6.80	6.65	6.49
26	7.87	7.72	7.57	7.42	7.27	7.12	6.97	6.82	6.67	6.52	6.37
27	7.73	7.58	7.44	7.29	7.14	6.99	6.84	6.69	6.55	6.40	6.25
28	7.60	7.45	7.30	7.16	7.01	6.86	6.71	6.57	6.42	6.27	6.12
29	7.47	7.32	7.17	7.03	6.88	6.73	6.59	6.44	6.29	6.15	6.00
30	7.34	7.19	7.05	6.90	6.75	6.61	6.46	6.31	6.17	6.02	5.87

PRESSURE(TORR)= 750

TEMP(C)   CHLORIDE(G/L)

	0	2	4	6	8	10	12	14	16	18	20
0	14.37	14.05	13.73	13.40	13.08	12.76	12.43	12.11	11.78	11.46	11.14
1	13.98	13.67	13.36	13.04	12.73	12.42	12.10	11.79	11.48	11.16	10.85
2	13.61	13.31	13.00	12.70	12.40	12.09	11.79	11.49	11.18	10.88	10.58
3	13.25	12.96	12.66	12.37	12.08	11.78	11.49	11.20	10.90	10.61	10.31
4	12.90	12.62	12.34	12.05	11.77	11.48	11.20	10.92	10.63	10.35	10.06
5	12.57	12.30	12.02	11.75	11.47	11.20	10.92	10.65	10.38	10.10	9.83
6	12.25	11.99	11.72	11.46	11.19	10.93	10.66	10.40	10.13	9.86	9.60
7	11.95	11.69	11.44	11.18	10.92	10.67	10.41	10.15	9.89	9.64	9.38
8	11.66	11.41	11.16	10.91	10.66	10.41	10.17	9.92	9.67	9.42	9.17
9	11.38	11.14	10.90	10.66	10.41	10.17	9.93	9.69	9.45	9.21	8.97
10	11.11	10.87	10.64	10.41	10.18	9.94	9.71	9.48	9.25	9.01	8.78
11	10.85	10.62	10.40	10.17	9.95	9.72	9.50	9.27	9.05	8.82	8.60
12	10.60	10.38	10.16	9.95	9.73	9.51	9.29	9.08	8.86	8.64	8.42
13	10.36	10.15	9.94	9.73	9.52	9.31	9.10	8.89	8.68	8.47	8.26
14	10.13	9.93	9.72	9.52	9.32	9.11	8.91	8.70	8.50	8.30	8.09
15	9.91	9.71	9.52	9.32	9.12	8.92	8.73	8.53	8.33	8.13	7.94
16	9.70	9.51	9.32	9.13	8.94	8.74	8.55	8.36	8.17	7.98	7.79
17	9.50	9.31	9.13	8.94	8.75	8.57	8.38	8.20	8.01	7.83	7.64
18	9.30	9.12	8.94	8.76	8.58	8.40	8.22	8.04	7.86	7.68	7.50
19	9.12	8.94	8.76	8.59	8.41	8.24	8.06	7.89	7.71	7.54	7.36
20	8.93	8.76	8.59	8.42	8.25	8.08	7.91	7.74	7.57	7.40	7.23
21	8.76	8.59	8.43	8.26	8.09	7.93	7.76	7.59	7.43	7.26	7.09
22	8.59	8.43	8.27	8.10	7.94	7.78	7.62	7.45	7.29	7.13	6.96
23	8.43	8.27	8.11	7.95	7.79	7.63	7.48	7.32	7.16	7.00	6.84
24	8.28	8.12	7.96	7.81	7.65	7.49	7.34	7.18	7.02	6.87	6.71
25	8.13	7.97	7.82	7.66	7.51	7.36	7.20	7.05	6.89	6.74	6.59
26	7.98	7.83	7.68	7.52	7.37	7.22	7.07	6.92	6.76	6.61	6.46
27	7.84	7.69	7.54	7.39	7.24	7.09	6.94	6.79	6.64	6.49	6.34
28	7.70	7.56	7.41	7.26	7.11	6.96	6.81	6.66	6.51	6.36	6.21
29	7.57	7.42	7.28	7.13	6.98	6.83	6.68	6.53	6.38	6.23	6.08
30	7.44	7.30	7.15	7.00	6.85	6.70	6.55	6.40	6.26	6.11	5.96

PRESSURE(TORR)= 760

TEMP(C)   CHLORIDE(G/L)

	0	2	4	6	8	10	12	14	16	18	20
0	14.57	14.24	13.91	13.58	13.26	12.93	12.60	12.27	11.94	11.61	11.29
1	14.17	13.85	13.54	13.22	12.90	12.58	12.27	11.95	11.63	11.31	11.00
2	13.79	13.48	13.18	12.87	12.56	12.26	11.95	11.64	11.33	11.03	10.72
3	13.43	13.13	12.83	12.54	12.24	11.94	11.64	11.35	11.05	10.75	10.45
4	13.08	12.79	12.50	12.21	11.93	11.64	11.35	11.06	10.78	10.49	10.20
5	12.74	12.46	12.19	11.91	11.63	11.35	11.07	10.79	10.52	10.24	9.96
6	12.42	12.15	11.88	11.61	11.34	11.07	10.80	10.54	10.27	10.00	9.73
7	12.11	11.85	11.59	11.33	11.07	10.81	10.55	10.29	10.03	9.77	9.51
8	11.81	11.56	11.31	11.06	10.81	10.55	10.30	10.05	9.80	9.55	9.30
9	11.53	11.29	11.04	10.80	10.56	10.31	10.07	9.82	9.58	9.34	9.09
10	11.26	11.02	10.79	10.55	10.31	10.08	9.84	9.61	9.37	9.14	8.90
11	10.99	10.77	10.54	10.31	10.08	9.85	9.63	9.40	9.17	8.94	8.72
12	10.74	10.52	10.30	10.08	9.86	9.64	9.42	9.20	8.98	8.76	8.54
13	10.50	10.29	10.07	9.86	9.65	9.43	9.22	9.01	8.79	8.58	8.37
14	10.27	10.06	9.86	9.65	9.44	9.24	9.03	8.82	8.62	8.41	8.20
15	10.05	9.85	9.65	9.45	9.25	9.05	8.85	8.65	8.45	8.24	8.04
16	9.83	9.64	9.44	9.25	9.06	8.86	8.67	8.47	8.28	8.09	7.89
17	9.63	9.44	9.25	9.06	8.87	8.69	8.50	8.31	8.12	7.93	7.74
18	9.43	9.25	9.06	8.88	8.70	8.52	8.33	8.15	7.97	7.78	7.60
19	9.24	9.06	8.88	8.71	8.53	8.35	8.17	7.99	7.82	7.64	7.46
20	9.06	8.88	8.71	8.54	8.36	8.19	8.02	7.84	7.67	7.50	7.32
21	8.88	8.71	8.54	8.37	8.20	8.04	7.87	7.70	7.53	7.36	7.19
22	8.71	8.55	8.38	8.22	8.05	7.89	7.72	7.56	7.39	7.22	7.06
23	8.55	8.39	8.22	8.06	7.90	7.74	7.58	7.42	7.25	7.09	6.93
24	8.39	8.23	8.07	7.91	7.76	7.60	7.44	7.28	7.12	6.96	6.80
25	8.24	8.08	7.93	7.77	7.61	7.46	7.30	7.14	6.99	6.83	6.68
26	8.09	7.94	7.78	7.63	7.47	7.32	7.17	7.01	6.86	6.70	6.55
27	7.95	7.80	7.64	7.49	7.34	7.19	7.03	6.88	6.73	6.58	6.42
28	7.81	7.66	7.51	7.36	7.21	7.05	6.90	6.75	6.60	6.45	6.30
29	7.68	7.53	7.38	7.23	7.07	6.92	6.77	6.62	6.47	6.32	6.17
30	7.55	7.40	7.25	7.10	6.95	6.79	6.64	6.49	6.34	6.19	6.04

PRESSURE(TORR)= 770

TEMP(C)   CHLORIDE(G/L)

	0	2	4	6	8	10	12	14	16	18	20
0	14.76	14.43	14.10	13.76	13.43	13.10	12.77	12.43	12.10	11.77	11.44
1	14.36	14.04	13.72	13.39	13.07	12.75	12.43	12.11	11.79	11.46	11.14
2	13.97	13.66	13.35	13.04	12.73	12.42	12.11	11.79	11.48	11.17	10.86
3	13.61	13.30	13.00	12.70	12.40	12.10	11.80	11.50	11.19	10.89	10.59
4	13.25	12.96	12.67	12.38	12.08	11.79	11.50	11.21	10.92	10.63	10.34
5	12.91	12.63	12.35	12.06	11.78	11.50	11.22	10.94	10.65	10.37	10.09
6	12.58	12.31	12.04	11.77	11.49	11.22	10.95	10.68	10.40	10.13	9.86
7	12.27	12.01	11.74	11.48	11.22	10.95	10.69	10.42	10.16	9.90	9.63
8	11.97	11.72	11.46	11.21	10.95	10.70	10.44	10.18	9.93	9.67	9.42
9	11.68	11.44	11.19	10.94	10.70	10.45	10.20	9.96	9.71	9.46	9.22
10	11.41	11.17	10.93	10.69	10.45	10.21	9.97	9.74	9.50	9.26	9.02
11	11.14	10.91	10.68	10.45	10.22	9.99	9.76	9.52	9.29	9.06	8.83
12	10.89	10.66	10.44	10.22	9.99	9.77	9.55	9.32	9.10	8.88	8.65
13	10.64	10.43	10.21	9.99	9.78	9.56	9.34	9.13	8.91	8.70	8.48
14	10.41	10.20	9.99	9.78	9.57	9.36	9.15	8.94	8.73	8.52	8.31
15	10.18	9.98	9.78	9.57	9.37	9.17	8.96	8.76	8.56	8.36	8.15
16	9.96	9.77	9.57	9.37	9.18	8.98	8.78	8.59	8.39	8.19	8.00
17	9.76	9.57	9.37	9.18	8.99	8.80	8.61	8.42	8.23	8.04	7.85
18	9.56	9.37	9.19	9.00	8.81	8.63	8.44	8.26	8.07	7.89	7.70
19	9.36	9.18	9.00	8.82	8.64	8.46	8.28	8.10	7.92	7.74	7.56
20	9.18	9.00	8.83	8.65	8.48	8.30	8.13	7.95	7.77	7.60	7.42
21	9.00	8.83	8.66	8.49	8.32	8.14	7.97	7.80	7.63	7.46	7.29
22	8.83	8.66	8.49	8.33	8.16	7.99	7.82	7.66	7.49	7.32	7.16
23	8.66	8.50	8.34	8.17	8.01	7.84	7.68	7.52	7.35	7.19	7.02
24	8.50	8.34	8.18	8.02	7.86	7.70	7.54	7.38	7.22	7.06	6.90
25	8.35	8.19	8.03	7.88	7.72	7.56	7.40	7.24	7.08	6.93	6.77
26	8.20	8.05	7.89	7.73	7.58	7.42	7.26	7.11	6.95	6.80	6.64
27	8.06	7.90	7.75	7.59	7.44	7.28	7.13	6.97	6.82	6.67	6.51
28	7.92	7.76	7.61	7.46	7.30	7.15	7.00	6.84	6.69	6.54	6.38
29	7.78	7.63	7.48	7.32	7.17	7.02	6.87	6.71	6.56	6.41	6.25
30	7.65	7.50	7.35	7.19	7.04	6.89	6.73	6.58	6.43	6.28	6.12

PRESSURE(TORR)= 780

TEMP(C)   CHLORIDE(G/L)

	0	2	4	6	8	10	12	14	16	18	20
0	14.95	14.62	14.28	13.94	13.61	13.27	12.93	12.60	12.26	11.92	11.59
1	14.55	14.22	13.90	13.57	13.24	12.92	12.59	12.27	11.94	11.61	11.29
2	14.16	13.84	13.53	13.21	12.90	12.58	12.26	11.95	11.63	11.32	11.00
3	13.78	13.48	13.17	12.87	12.56	12.26	11.95	11.65	11.34	11.04	10.73
4	13.42	13.13	12.83	12.54	12.24	11.95	11.65	11.36	11.06	10.77	10.47
5	13.08	12.79	12.51	12.22	11.94	11.65	11.37	11.08	10.79	10.51	10.22
6	12.75	12.47	12.20	11.92	11.64	11.37	11.09	10.82	10.54	10.26	9.99
7	12.43	12.17	11.90	11.63	11.36	11.10	10.83	10.56	10.29	10.03	9.76
8	12.13	11.87	11.61	11.35	11.09	10.84	10.58	10.32	10.06	9.80	9.54
9	11.84	11.59	11.34	11.09	10.84	10.59	10.34	10.09	9.84	9.59	9.34
10	11.56	11.31	11.07	10.83	10.59	10.35	10.11	9.86	9.62	9.38	9.14
11	11.29	11.05	10.82	10.59	10.35	10.12	9.88	9.65	9.42	9.18	8.95
12	11.03	10.80	10.58	10.35	10.12	9.90	9.67	9.44	9.22	8.99	8.77
13	10.78	10.56	10.34	10.12	9.91	9.69	9.47	9.25	9.03	8.81	8.59
14	10.54	10.33	10.12	9.91	9.70	9.48	9.27	9.06	8.85	8.63	8.42
15	10.32	10.11	9.90	9.70	9.49	9.29	9.08	8.88	8.67	8.47	8.26
16	10.10	9.90	9.70	9.50	9.30	9.10	8.90	8.70	8.50	8.30	8.10
17	9.89	9.69	9.50	9.31	9.11	8.92	8.73	8.53	8.34	8.15	7.95
18	9.68	9.50	9.31	9.12	8.93	8.74	8.56	8.37	8.18	7.99	7.80
19	9.49	9.31	9.12	8.94	8.76	8.57	8.39	8.21	8.03	7.84	7.66
20	9.30	9.12	8.95	8.77	8.59	8.41	8.23	8.06	7.88	7.70	7.52
21	9.12	8.95	8.77	8.60	8.43	8.25	8.08	7.91	7.73	7.56	7.38
22	8.95	8.78	8.61	8.44	8.27	8.10	7.93	7.76	7.59	7.42	7.25
23	8.78	8.61	8.45	8.28	8.11	7.95	7.78	7.62	7.45	7.28	7.12
24	8.62	8.45	8.29	8.13	7.97	7.80	7.64	7.48	7.31	7.15	6.99
25	8.46	8.30	8.14	7.98	7.82	7.66	7.50	7.34	7.18	7.02	6.86
26	8.31	8.15	7.99	7.84	7.68	7.52	7.36	7.20	7.04	6.89	6.73
27	8.17	8.01	7.85	7.70	7.54	7.38	7.23	7.07	6.91	6.76	6.60
28	8.02	7.87	7.71	7.56	7.40	7.25	7.09	6.94	6.78	6.62	6.47
29	7.89	7.73	7.58	7.42	7.27	7.11	6.96	6.80	6.65	6.49	6.34
30	7.76	7.60	7.45	7.29	7.14	6.98	6.83	6.67	6.52	6.36	6.21

PRESSURE(TORR)= 790

TEMP(C)   CHLORIDE(G/L)

	0	2	4	6	8	10	12	14	16	18	20
0	15.15	14.80	14.46	14.12	13.78	13.44	13.10	12.76	12.42	12.08	11.74
1	14.73	14.40	14.07	13.74	13.41	13.08	12.75	12.42	12.09	11.76	11.43
2	14.34	14.02	13.70	13.38	13.06	12.74	12.42	12.10	11.78	11.46	11.14
3	13.96	13.65	13.34	13.03	12.72	12.42	12.11	11.80	11.49	11.18	10.87
4	13.60	13.30	13.00	12.70	12.40	12.10	11.80	11.50	11.20	10.91	10.61
5	13.25	12.96	12.67	12.38	12.09	11.80	11.51	11.22	10.93	10.64	10.36
6	12.91	12.63	12.35	12.07	11.79	11.51	11.23	10.95	10.68	10.40	10.12
7	12.59	12.32	12.05	11.78	11.51	11.24	10.97	10.70	10.43	10.16	9.89
8	12.29	12.02	11.76	11.50	11.24	10.98	10.71	10.45	10.19	9.93	9.67
9	11.99	11.74	11.48	11.23	10.98	10.72	10.47	10.22	9.96	9.71	9.46
10	11.71	11.46	11.22	10.97	10.73	10.48	10.24	9.99	9.75	9.50	9.26
11	11.43	11.20	10.96	10.72	10.49	10.25	10.01	9.78	9.54	9.30	9.06
12	11.17	10.94	10.71	10.48	10.26	10.03	9.80	9.57	9.34	9.11	8.88
13	10.92	10.70	10.48	10.26	10.03	9.81	9.59	9.37	9.15	8.92	8.70
14	10.68	10.47	10.25	10.04	9.82	9.61	9.39	9.18	8.96	8.75	8.53
15	10.45	10.24	10.03	9.83	9.62	9.41	9.20	8.99	8.78	8.58	8.37
16	10.23	10.03	9.82	9.62	9.42	9.22	9.02	8.81	8.61	8.41	8.21
17	10.01	9.82	9.62	9.43	9.23	9.04	8.84	8.64	8.45	8.25	8.06
18	9.81	9.62	9.43	9.24	9.05	8.86	8.67	8.48	8.29	8.10	7.91
19	9.61	9.43	9.24	9.06	8.87	8.69	8.50	8.32	8.13	7.95	7.76
20	9.42	9.24	9.06	8.88	8.70	8.52	8.34	8.16	7.98	7.80	7.62
21	9.24	9.06	8.89	8.71	8.54	8.36	8.19	8.01	7.83	7.66	7.48
22	9.06	8.89	8.72	8.55	8.38	8.21	8.03	7.86	7.69	7.52	7.35
23	8.90	8.73	8.56	8.39	8.22	8.05	7.89	7.72	7.55	7.38	7.21
24	8.73	8.57	8.40	8.24	8.07	7.91	7.74	7.58	7.41	7.24	7.08
25	8.57	8.41	8.25	8.09	7.92	7.76	7.60	7.44	7.27	7.11	6.95
26	8.42	8.26	8.10	7.94	7.78	7.62	7.46	7.30	7.14	6.98	6.82
27	8.27	8.12	7.96	7.80	7.64	7.48	7.32	7.16	7.00	6.85	6.69
28	8.13	7.97	7.82	7.66	7.50	7.34	7.19	7.03	6.87	6.71	6.56
29	7.99	7.84	7.68	7.52	7.37	7.21	7.05	6.89	6.74	6.58	6.42
30	7.86	7.70	7.55	7.39	7.23	7.07	6.92	6.76	6.60	6.45	6.29

PRESSURE(TORR)= 800

TEMP(C)   CHLORIDE(G/L)

	0	2	4	6	8	10	12	14	16	18	20
0	15.34	14.99	14.65	14.30	13.96	13.61	13.27	12.92	12.58	12.23	11.88
1	14.92	14.59	14.25	13.92	13.58	13.25	12.92	12.58	12.25	11.91	11.58
2	14.52	14.20	13.88	13.55	13.23	12.90	12.58	12.26	11.93	11.61	11.29
3	14.14	13.83	13.51	13.20	12.89	12.57	12.26	11.95	11.63	11.32	11.01
4	13.77	13.47	13.17	12.86	12.56	12.26	11.95	11.65	11.35	11.04	10.74
5	13.42	13.13	12.83	12.54	12.25	11.95	11.66	11.37	11.07	10.78	10.49
6	13.08	12.80	12.51	12.23	11.95	11.66	11.38	11.09	10.81	10.53	10.24
7	12.75	12.48	12.21	11.93	11.66	11.38	11.11	10.83	10.56	10.29	10.01
8	12.44	12.18	11.91	11.65	11.38	11.12	10.85	10.59	10.32	10.06	9.79
9	12.14	11.89	11.63	11.37	11.12	10.86	10.60	10.35	10.09	9.83	9.58
10	11.86	11.61	11.36	11.11	10.86	10.62	10.37	10.12	9.87	9.62	9.37
11	11.58	11.34	11.10	10.86	10.62	10.38	10.14	9.90	9.66	9.42	9.18
12	11.32	11.08	10.85	10.62	10.39	10.15	9.92	9.69	9.46	9.23	8.99
13	11.06	10.84	10.61	10.39	10.16	9.94	9.71	9.49	9.26	9.04	8.81
14	10.82	10.60	10.38	10.17	9.95	9.73	9.51	9.29	9.08	8.86	8.64
15	10.58	10.37	10.16	9.95	9.74	9.53	9.32	9.11	8.90	8.69	8.48
16	10.36	10.16	9.95	9.75	9.54	9.34	9.13	8.93	8.72	8.52	8.31
17	10.14	9.95	9.75	9.55	9.35	9.15	8.95	8.75	8.56	8.36	8.16
18	9.94	9.74	9.55	9.36	9.17	8.97	8.78	8.59	8.39	8.20	8.01
19	9.74	9.55	9.36	9.17	8.99	8.80	8.61	8.42	8.24	8.05	7.86
20	9.54	9.36	9.18	9.00	8.81	8.63	8.45	8.27	8.08	7.90	7.72
21	9.36	9.18	9.00	8.83	8.65	8.47	8.29	8.11	7.94	7.76	7.58
22	9.18	9.01	8.83	8.66	8.49	8.31	8.14	7.96	7.79	7.62	7.44
23	9.01	8.84	8.67	8.50	8.33	8.16	7.99	7.82	7.65	7.48	7.31
24	8.85	8.68	8.51	8.34	8.18	8.01	7.84	7.67	7.51	7.34	7.17
25	8.69	8.52	8.36	8.19	8.03	7.86	7.70	7.53	7.37	7.20	7.04
26	8.53	8.37	8.21	8.04	7.88	7.72	7.56	7.39	7.23	7.07	6.91
27	8.38	8.22	8.06	7.90	7.74	7.58	7.42	7.26	7.10	6.93	6.77
28	8.24	8.08	7.92	7.76	7.60	7.44	7.28	7.12	6.96	6.80	6.64
29	8.10	7.94	7.78	7.62	7.46	7.30	7.14	6.98	6.83	6.67	6.51
30	7.96	7.80	7.64	7.49	7.33	7.17	7.01	6.85	6.69	6.53	6.37

INTENSIVE AND EXTENSIVE PROPERTIES

In considering the thermodynamic aspects of oxygen solubility it is useful to note the two classes of properties which are measured quantitatively. If we consider two identical systems, say two kilogram weights of steel, the mass of the two is obviously double that of each one. Properties such as this in which the total value of a system is the sum of the individual values for each of its constituent parts are called *extensive*. On the other hand, the temperature of two identical pieces of steel is the same as that of either one, just as, say, the density is. Properties of this type which are not additive and which can be specified for any system without reference to the size of that system are called *intensive*. Examples of extensive and intensive properties for various systems are given in Table C.1 (1).

TABLE C.1 Extensive and Intensive Properties

Type of system	Intensive property	Extensive property	Work done by or on system
Gravitational	Height (h)	Mass (m)	$mg \, \Delta h$
Thermal	Temperature (θ)	Heat capacity (C)	$C \, \Delta\theta$
Electrical	Voltage (E)	Charge (q)	qE
Chemical	Chemical potential (μ)	Number of moles (N)	$N \, \Delta\mu$

The use of the term potential for chemical systems is no different in principle from its more familiar use in mechanical and electrical systems. In these more familiar systems the potential has associated with it a capacity quantity, and the work done on the system is the product of this capacity factor and the change in potential. So for gravitational potential, $g \, \Delta h$, the mass is the capacity factor and the work done is $mg \, \Delta h$. And for the electrostatic system the charge q is the capacity factor, and the work done in moving the charge from one location to another, where the potential difference between the two is ΔE, is $q \, \Delta E$.

With chemical systems using the term chemical potential and defining the capacity factor as the mole, then to transfer N moles of substance between two points differing in chemical potential by $\Delta\mu$ requires an amount of work equal to $N\,\Delta\mu$. And just as a mass or charge will move spontaneously to the state of lowest gravitational or electrostatic potential, so a chemical substance, if it is free to move, will move spontaneously to a lower chemical potential. In other words, the chemical potential is just another measure of the escaping tendency of a substance in a system. At equilibrium the chemical potential must be constant throughout the entire system for each substance that is free to move.

Since both the chemical potential and fugacity are measures of escaping tendency, then they should be related, and eq. (C.1) shows this relationship (2, p. 262):

$$\mu_i = \mu^* + RT \ln f_i \tag{C.1}$$

where μ^* can be regarded either as the chemical potential of a real gas when its fugacity is unity, or as the chemical potential of the same gas if it behaved ideally at unit (1 atm) partial pressure; the second viewpoint follows from what we have said about the relationship between fugacity and partial pressure [Section 2.1]. Also, since the fugacity is directly related to the activity [eq. (2.18)] we can express the chemical potential in terms of activity:

$$\mu_i = \mu^0 + RT \ln a_i \tag{C.2}$$

where μ^0 is the chemical potential corresponding to unit activity. As is discussed in Section 2.4, the state of unit activity is the standard state and there is some freedom in the choice of this state. If we choose the standard state of unit activity as that in which the fugacity of the gas is unity, then this choice makes μ^0 in eq. (C.2) identical with μ^* in eq. (C.1). Or, looking at it another way, the proportionality constant, k, in eq. (2.18) is equal to unity.

However, whether or not this coincidence of fugacity and activity occurs, eq. (C.2) shows that the activity depends solely on the difference in chemical potential of any state and that of the chosen standard state, whatever that may be. And since the chemical potential, or escaping tendency, does not depend on the amount of the system one takes, but only on the extent to which the system is removed from equilibrium, the activity is therefore an intensive property. This is in contrast to the number of moles in the system, which is an extensive property of the system.

For chemical systems it is important to distinguish between intensive properties based on measuring the chemical potential, and extensive properties based on measuring the actual number of moles of a given substance. This can be illustrated if we compare a potentiometric pH measurement, say, with a measurement of the number of H^+ ions by a titrimetric method.

In the case of pH the measurement is based on a potential determination, which is essentially an intensive property. But for the determination of H^+

by titration the process is based on a stoichiometric calculation, and so an extensive property is being measured. The results of the two methods of analysis may not agree, particularly if interferences are present which may cause the activity coefficient, γ, which relates activity and concentration, to deviate from unity; for example, when salting-out occurs.

These considerations become especially important for polarographic oxygen measurements. In such determinations the measured property is an intensive one since the current of the detector is dependent only on the difference in the chemical potential of oxygen across the membrane. Thus, results obtained by polarographic means do not necessarily have to agree with results obtained for dissolved oxygen by titration methods (e.g. the Winkler test). Polarographic detectors measure the activity of molecular oxygen, while titrations measure the total number of oxygen molecules present. In the absence of complications these can be the same, but when salting-out occurs, for example, the activity of dissolved oxygen barely changes, while the concentration of the gas can fall dramatically [cf. Fig. 5.22]. By contrast to this effect of salting-out on polarographic detectors, when the reading obtained remains more or less constant even though the oxygen concentration is falling as the salt concentration increases, measurements made under high hydrostatic pressure will increase as the depth increases, even though the oxygen solubility remains effectively constant [Sections 2.6 and 5.3.4]. Therefore, caution is necessary under such conditions as these; but at the same time it should be noted that the ability to measure the chemical potential of oxygen directly is extremely important for it is the chemical potential that is the "effective concentration" in solution. Thus it is the activity and not the concentration that determines the rate of corrosion of metals, or the oxygen balance in a river or lake, or the quantity of oxygen transferable across a membrane. Those working in the life sciences clearly recognize this fact by referring to dissolved oxygen in terms of partial pressure, p_{O_2} (or oxygen tension), which is directly related to the chemical potential [eq. (C.1)].

REFERENCES

1. K. H. Mancy, *Instrumental Analysis for Water Pollution Control*, Ann Arbor Science Publishers Inc., Ann Arbor, Mich., 1971.

2. S. Glasstone, *Thermodynamics for Chemists*, D. Van Nostrand Company, New Jersey, 1947.

D

DIMENSIONAL ANALYSIS

Any physical equation must be physically homogeneous—that is, the units on both sides of the equality sign must be the same—and any equation derived to describe a given system should be checked for dimensional homogeneity. Conversely, since an equation must be homogeneous then often a nonrigorous derivation of the form of an equation describing a physical system can be obtained from very simple considerations.

Consider the case of the time-dependent diffusion layer thickness, δ_t, when only diffusional transport is occurring. In Section 3.6.2 it was guessed that δ_t was a function only of the diffusion coefficient and time:

$$\delta_t = f(D, t) \tag{3.34}$$

This can be written as an equality if we introduce a proportionality constant K, and powers of D and t, ε and ζ, respectively:

$$\delta_t = KD^\varepsilon t^\zeta \tag{3.35}$$

Expressing this in dimensional units,

$$\text{Length} = K \, (\text{length})^{2\varepsilon} \, (\text{time})^{-\varepsilon} \, (\text{time})^\zeta$$

and for dimensional homogeneity,

$$2\varepsilon = 1$$

and

$$\varepsilon - \zeta = 0$$

Hence $\varepsilon = \zeta = \frac{1}{2}$ and

$$\delta_t = K(Dt)^{1/2} \tag{3.36}$$

Dimensional analysis can never give the value of the constant, K, but nevertheless it is a useful tool to gain insight into a physical situation.

It should be noted that dimensional analysis of this type will not give a complete solution when dimensionless groups appear in the equation. Thus although one can write for the diffusion layer in convective diffusion on intuitive grounds,

$$\delta = f(D, v, v_m, l)$$

214

this cannot be solved since D and v are related by the dimensionless Schmidt number [eq. (G.1)] and v, v_m, and l by the dimensionless Reynolds' number [eq. (3.19)]. Rewriting eq. (3.26) in this form gives

$$\delta = \frac{1}{(\mathrm{Re})^{1/2}(\mathrm{Sc})^{1/2}}$$

and with such instances as this a more detailed analysis is necessary.

THE TRANSFER COEFFICIENT

When any substance undergoes a chemical reaction it has to overcome an energy barrier, in much the same way that a mechanical process has to. Figure E.1 illustrates this concept for the mechanical case. Here a box resting initially in an upright position on a flat surface (position A) is tilted about one edge until it is lying flat on the surface (position C). It can be seen that during the displacement the center of gravity of the box has passed through a maximum (B) which corresponds to a maximum in the potential energy. Similar principles to this simple mechanical model apply to more complex physicochemical systems, such as chemical and electrochemical reactions. Let us first consider the case of a charge transfer reaction in the absence of an electrical field; in other words, a chemical reaction.

Figure E.2 shows the progress of a reaction in terms of the energy changes occurring during the reaction. The reactants start off in a position of minimum energy and the products end up also in an energy minimum, but in between, much the same as in the mechanical case, there is a region with an energy maximum which constitutes an energy barrier to the reaction. The final state is intrinsically more stable than the initial state, being of lower energy, and from a thermodynamic point of view the reaction should occur spontaneously since the system as a whole will tend to move to a position of minimum energy. However, before charge transfer can be accomplished

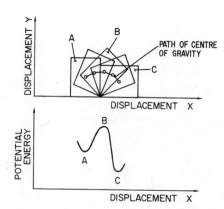

Fig. E.1 Potential energy profile for a mechanical system.

216

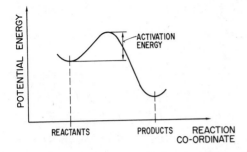

Fig. E.2 Potential energy profile for a chemical system.

the reactants have to have a certain *activation energy*; this brings in kinetics since the frequency with which they gain this energy in order to overcome the energy barrier has to be considered. The theoretical derivation of the rate of a reaction is beyond the scope of this account, but it is clear that since it involves the energies of the initial, final, and *activated* or *transition state*, and the frequency with which the reactants achieve sufficient energy to reach the transition state, then a knowledge of the form of the transition state is required.

For chemical kinetics such a model often exists, but for electrochemical kinetics the situation is complicated by the presence of the electrical field, which means, among other things, that the activation energy is made up of a potential independent part—chemical—and a potential dependent part—electrochemical—and generally a detailed model is not available. Therefore, in order to be able to calculate the reaction rate it is assumed that the potential dependent part of the activation energy is some fraction α of the potential dependent part of the difference in energy between the initial state before charge transfer and the final state after transfer. The fraction α for a simple one-electron transfer reaction is generally about 0.5 (1, Ch. 8). Further discussion of the significance of α for electrochemical reactions can be found in a number of places, for example (1–3).

REFERENCES

1. J. O'M. Bockris and A. K. N. Reddy, *Modern Electrochemistry*, Plenum Press, New York, 1970.

2. J. O'M. Bockris and Z. Nagy, "Symmetry factor and transfer coefficient," *J. Chem. Ed.*, **50**, 839 (1973).

3. P. Delahay, *Double Layer and Electrode Kinetics*, Wiley, New York, 1965.

F

CONVECTIVE FLOW

Here we expand a little on the discussion about convection in Section 3.3.2. Let us look at what happens, for example, when a fluid flows over a surface such as AB—Fig. F.1. It is found experimentally that a layer D at a distance $(x + dx)$ from AB flows with a velocity greater than that of a layer C at a distance x from AB. If the difference in the velocities of the two layers is dv the velocity gradient between C and D will be dv/dx. As a result of this relative motion of the layers internal friction or *viscosity* arises. This viscous nature of a fluid is only shown when one region of the fluid moves relative to another. If the layers moving relative to one another are parallel then the flow is said to be *streamlined* or *laminar*; otherwise the flow is disorderly or *turbulent*.

For streamlined motion Newton showed that the tangential viscous force between two layers of area A cm^2, a distance dx apart and moving with relative velocity dv, is given by

$$F = \eta A \frac{dv}{dx} \tag{F.1}$$

or

$$f = \eta \frac{dv}{dx} \tag{F.2}$$

where

$$f = \frac{F}{A}$$

The proportionality constant, η, is a characteristic constant of the fluid and is called the *coefficient of viscosity*. Dimensional analysis [Appendix D]

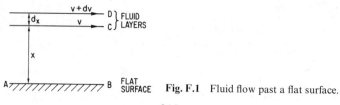

Fig. F.1   Fluid flow past a flat surface.

**TABLE F.1   Values of Coefficient of Viscosity
and of Kinematic Viscosity**

Substance	$\eta\,(\mathrm{g\,cm^{-1}\,s^{-1}})$ at 20°C	$\nu\,(\mathrm{cm^2\,s^{-1}})$
Water	1.0×10^{-2}	1.0×10^{-2}
Air	1.8×10^{-4}	1.5×10^{-1}
Mercury	1.6×10^{-2}	1.2×10^{-3}
Glycerine	8.5	6.8

shows that the units of η are g cm^{-1} s^{-1}. The values of η for different fluids show a very wide spread—Table F.1.

The variation of viscosity with temperature is considerable and can be approximately described by a classical activation equation,

$$\eta = \eta_0 \exp\left(\frac{E_v}{RT}\right) \tag{F.3}$$

with η_0 being a constant, E_v the energy barrier to viscous flow, and R the gas constant. Table F.2 gives values of the viscosity of water at various temperatures.

**TABLE F.2   Temperature Variation of
Viscosity and Kinematic Viscosity for Water**

Temp. (°C)	$\eta\,(\mathrm{g\,cm^{-1}\,s^{-1}} \times 10^2)$	$\nu\,(\mathrm{cm^2\,s^{-1}} \times 10^2)$
0	1.787	1.787
20	1.002	1.004
40	0.653	0.658
60	0.467	0.475
80	0.355	0.365
100	0.282	0.294

Very often in dealing with fluid dynamic problems the ratio of the coefficient of viscosity and the density of the fluid occurs. This ratio is called the *kinematic viscosity*:

$$\nu = \frac{\eta}{\rho} \tag{F.4}$$

and has units cm^2 s^{-1}; typical values are given in Table F.1. The temperature dependence of ν is obviously a combined function of the temperature variation of η and ρ, but for water or aqueous solutions ρ varies only slightly and so v follows η closely—Table F.2.

The quantities which are characteristic of a moving liquid can be represented by a function of a single controlling dimensionless number—the *Reynolds' number*, Re. This number is defined as

$$Re = \frac{v_c l}{v} \tag{F.5}$$

where v_c is the fluid velocity at which there is a change from laminar flow to turbulent flow—the *critical velocity*; the characteristic length of the body is l and v is the kinematic viscosity. For aqueous solutions streamline flow is present for $Re \ll \sim 10^3$ and turbulence for $Re \gg \sim 10^3$, which means that with $l \sim 1$ cm and $v \sim 10^{-2}$ cm^2 s^{-1}, $v_c \sim 10$ cm s^{-1}.

G

CONVECTIVE DIFFUSION

This appendix amplifies the discussion of Section 3.3.4 on convective diffusion.

The ratio of the two dimensionless numbers Re, for convection, and Pe, for convective diffusion, is known as the Schmidt number, Sc:

$$Sc = \frac{Pe}{Re} = \frac{vl}{D} \frac{v}{vl} = \frac{v}{D} \tag{G.1}$$

The Schmidt number is not a function of flow velocity or of characteristic dimensions, but is determined exclusively by the physical properties which characterize momentum transfer—that is, viscosity and density—and transport of matter by a purely molecular mechanism—that is, diffusion.

For liquids, such as water, $v \sim 10^{-2}$ cm^2 s^{-1} and $D \sim 10^{-5}$ cm^2 s^{-1}, and so Sc $\sim 10^3$. With this value we see that Pe is $\gg 1$ for Re as low as 10^{-2}, and so for values of the Reynolds' number met in practice ($\sim 10^3$) Pe is very high. Thus even for low values of Re convective transfer is more important than diffusion. Looking at it another way, as v increases so D decreases approximately according to the relation

$$D = \frac{constant}{v} \tag{G.2}$$

So Sc increases as the square of v, and in viscous fluids Sc can be $\sim 10^6$ or higher. A high value of Sc indicates that even at very low velocities the transport of matter in a fluid by convection predominates over molecular diffusion.

VARIATION OF THE STEADY-STATE CURRENT AT A MPOD WITH ELECTROLYTE LAYER THICKNESS

In Chapter 4 the steady-state current for a MPOD was derived assuming that the characteristic time for oxygen diffusion in the membrane is very much longer than the corresponding time of diffusion in the electrolyte layer. However, the current is in fact dependent on the thickness of the electrolyte layer, but this dependence cannot be very readily seen if, as in Chapter 4, the expression for the steady-state current is derived from the transient current obtained on switching on the detector because of the complex nature of the time-dependent solution. If, on the other hand, one considers just the steady-state situation for oxygen diffusion through a two-layer system then the dependence becomes quite clear.

The system is that illustrated in Fig. 4.4 for t_∞. A linear concentration gradient in each phase (membrane and electrolyte) is assumed, and we consider only the case where the oxygen concentration at the electrode surface is zero—that is, the maximum flux condition [cf. Section 3.5]. The steady-state oxygen flux, J_m, through the membrane of thickness b is given by eq. (4.9)

$$J_m = D_m \frac{C_{m, x=b} - C_{m, x=0}}{b} \qquad (4.9)$$

where D_m is the diffusion coefficient of oxygen in the membrane, and $C_{m, x=b}$ and $C_{m, x=0}$ are the oxygen concentrations within the membrane at the test solution and electrolyte film interfaces, respectively. Using the relationships of eqs. (4.2) and (4.3) gives

$$J_m = D_m \frac{K_b C_s - K_0 C_{e, x=0}}{b} \qquad (H.1)$$

where K_b and K_0 are the distribution coefficients at the two interfaces, and C_s and $C_{e, x=0}$ are the concentrations in the test solution and electrolyte film, also at the interfaces. The flux of oxygen, J_e, through the electrolyte of thickness a is readily obtained from eq. (3.27):

$$J_e = D_e \frac{C_{e, x=0}}{a} \qquad (H.2)$$

222

where D_e is the oxygen diffusion coefficient in the electrolyte film. Since in the steady state $J_e = J_m$ it follows that the concentration of oxygen in the electrolyte at the interior surface of the membrane is related to the concentration in the test solution at the outer surface by

$$C_{e, x=0} = \frac{K_b}{K_0 + \dfrac{D_e b}{D_m a}} C_s \qquad (H.3)$$

The current at an electrode is directly proportional to the flux of electroactive material to the electrode [Section 3.5], and so for a MPOD the expression for the current density, which we write as [cf. eq. (3.29)]

$$i_{L, a \neq 0} = nF J_e = nF J_m \qquad (H.4)$$

becomes

$$i_{L, a \neq 0} = nF \frac{K_b D_m}{b} C_s \frac{1}{1 + K_0 \dfrac{D_m a}{D_e b}} \qquad (H.5)$$

where $i_{L, a \neq 0}$ is the current density corresponding to nonzero thickness of the electrolyte layer. Using eq. (4.11) to relate D_m to the permeability coefficient, P_m, and eq. (4.17),

$$i_L = nF \frac{P_m}{b} C_s \qquad (4.17)$$

it is seen that $i_{L, a \neq 0}$ can be expressed in terms of the current for an electrolyte layer of zero thickness by

$$i_{L, a \neq 0} = \frac{i_L}{1 + K_0 \dfrac{D_m a}{D_e b}} \qquad (H.6)$$

Clearly when $a = 0$, $i_{L, a \neq 0} = i_L$ and so i_L represents a maximum current for a given membrane.

The variation of the steady-state current with the thickness of the electrolyte layer is most conveniently analyzed by rearranging eq. (H.6) to

$$\frac{1}{i_{L, a \neq 0}} = \frac{1}{i_L} + \frac{1}{i_L} K_0 \frac{D_m a}{D_e b} \qquad (H.7)$$

Figure H.1 shows a plot of eq. (H.7) for a MPOD with different electrolyte layer thicknesses (1); the varying thickness is conveniently achieved by means of different size spacers underneath the membrane and the distance a is measured with a light-splitting microscope. The linear relationship

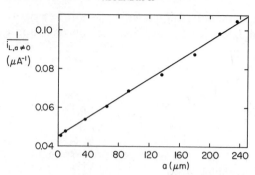

Fig. H.1   Test of eq. (H.7). Results obtained with an Orbisphere Model 2101 oxygen indicator with $1M$ KOH electrolyte and a 12.5 μm Teflon FEP membrane; temperature—23°C; cathode radius—0.316 cm (1).

predicted by eq. (H.7) is clearly obtained, and the plot can be used to determine the transport parameters of oxygen in the system. The intercept for zero thickness of the electrolyte gives i_L, which allows a more precise determination of the permeability coefficient, P_m, than given by eq. (4.17), which assumes $a = 0$ but which is in fact for $a \sim 10$ μm; calculating P_m directly from eq. (4.17) gives a value about 5% too low. For the given conditions P_m is calculated from the intercept to be 8.7×10^{-7} cm^2 s^{-1}—the value calculated directly from eq. (4.17) is 8.2×10^{-7} [Section 5.1].

The slope of the straight line reflects the influence of the transport through the electrolyte on the steady-state flux of oxygen from the test solution to the cathode. From Fig. H.1 the gradient of the plot is found to be 2.48×10^6 A^{-1} cm^{-1}. From the literature (2, 3) the ratio of the solubility of oxygen in $1M$ KOH to that in water is 0.664, and from eq. (4.8) this means

$$\frac{K_b}{K_0} = 0.664$$

Thus

$$K_0 D_m = \frac{K_b D_m}{0.664} = \frac{P_m}{0.664}$$

Therefore the slope is related to D_e by

$$2.48 \times 10^6 = \frac{1}{i_L} \frac{P_m}{0.664 D_e b}$$

Substituting the value of P_m obtained from the intercept together with the values of i_L and membrane thickness, b, allows the diffusion coefficient for oxygen in $1M$ KOH to be calculated:

$$D_e = 1.90 \times 10^{-5} \text{ cm}^2 \text{ s}^{-1}$$

REFERENCES

1. J. M. Hale, personal communication.
2. S. K. Shoor, R. D. Walker, and K. E. Gubbins, "Salting out of non-polar gases in aqueous potassium hydroxide solutions," *J. Phys. Chem.*, **73**, 312 (1969).
3. E. W. Tiepel and K. E. Gubbins, "Thermodynamic properties of gases dissolved in electrolyte solutions," *Ind. Eng. Chem. Fundam.*, **12**, 18 (1973).

CALCULATION OF MEMBRANE PERMEABILITIES

Figure I.1 shows the comprehensive diagram for the estimation of the diffusion coefficients of gases in membranes (1, Ch. 18). Taking the case of oxygen diffusing in PTFE as an example, one starts in the upper-left section at the glass transition temperature for PTFE $- T_g = 400$ K. From this point we move vertically until we meet curve b (glassy polymer for PTFE), and then from there we go horizontally to the upper-right section where the O_2 line is met. We then descend to read the activation energy for diffusion [Section 3.3.3] as $E_D \sim 7.5$ kcal mol^{-1}. Further down we meet line 2 from where we move horizontally to the point corresponding to $E_D \sim 7.5$ kcal mol^{-1}. Going upward from this point gives log $D(298) = -6.6$ or $D = 2.5 \times 10^{-7}$ cm^2 s^{-1}. This value is, however, strictly a value for an amorphous polymer, and since PTFE is a semicrystalline material the value has to be corrected:

$$D_{sc} = D_a(1 - x_c) \tag{I.1}$$

where x_c is the degree of crystallinity. Typically $x_c \sim 0.6$ and so $D_{sc} \sim 1.0 \times 10^{-7}$ cm^2 s^{-1}.

The solubility, $S_{a,e}(298)$, of a gas in an amorphous elastomer at 298 K is given by, to a first approximation,

$$\log S_{a,e}(298) \simeq -2.1 + 0.0123T_b \tag{I.2}$$

where T_b is the boiling point of the gas. For oxygen, $T_b = 90.2$ K and so $S_{a,e}(298) \sim 0.1$ cm^3 (cm^3 of polymer)$^{-1}$ atm^{-1}. But this solubility is for the gas in an elastomeric material and we require the solubility in a material which is below the glass transition temperature—that is, in a glassy state. To calculate this solubility we first obtain the heat of solution, ΔH_r, of the gas in the rubbery state of the polymer. This is described by the following expression:

$$\Delta H_r = -2400 - 2000 \log S_{a,e}(298) \tag{I.3}$$

which in our case we calculate to be -419 cal mol^{-1}, or -1.75 kJ mol^{-1}. The solubility, $S_{a,e}(T_g)$, at the glass transition point, T_g, is next calculated using the relationship

$$S_{a,e}(T_g) = S_{a,e}(298) \exp\left[\frac{\Delta H_r}{R}\left(\frac{1}{298} - \frac{1}{T_g}\right)\right] \tag{I.4}$$

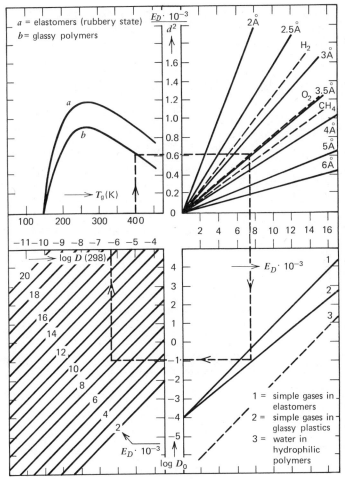

Fig. I.1 Diagram for the estimation of the diffusion coefficient of a gas in a polymer (1). (Reproduced by permission of the Elsevier Publishing Company.)

and for $T_g = 400$ K this gives $S_{a,e}(T_g) = 0.083$. Now the heat of solution, ΔH_g, in a glassy polymer is approximately related to ΔH_r by

$$\Delta H_g \simeq 1.67\, \Delta H_r - 5000 \qquad (I.5)$$

and so $\Delta H_g = -23.9$ kJ mol^{-1}. The solubility at T_g can now in turn be related to this heat of solution by

$$S_{a,e}(T_g) = S_0 \exp\left(-\frac{\Delta H_g}{RT_g}\right) \qquad (I.6)$$

where S_0 is a preexponential factor. Using the value of $S_{a,e}(T_g)$ found with eq. (I.4), S_0 is calculated as 6.36×10^{-5}. Finally, with S_0 the gas solubility at 298 K in a glassy, amorphous polymer is evaluated using eq. (I.6):

$$S_{a,g}(298) = S_0 \exp\left(-\frac{\Delta H_g}{298R}\right)$$

$$= 0.98 \text{ cm}^3 \, (\text{cm}^3 \text{ of polymer})^{-1} \, \text{atm}^{-1}$$

Correcting for the degree of crystallinity again,

$$S_{sc,g} = S_{a,g}(1 - x_c)$$

gives $S_{sc,g} \sim 0.4 \text{ cm}^3 \, (\text{cm}^3 \text{ of polymer})^{-1} \, \text{atm}^{-1}$. With a partial pressure of ~ 0.2 atm for oxygen dissolved in air-saturated water, $S_{sc,g}$ converts to $1.1 \times 10^{-4} \text{ g } (\text{cm}^3 \text{ of polymer})^{-1}$ or 110 mg liter^{-1}. At 25°C the oxygen solubility in water is 8 mg liter^{-1} and therefore we have a value of $K_b \sim 13.7$. Using eq. (4.11) the permeability coefficient, P_m, can be calculated from this value of K_b and the value of D_{sc} obtained from eq. (I.1), and it is found to be $\sim 14 \times 10^{-7} \text{ cm}^2 \text{ s}^{-1}$.

The values of D, P_m, and K_b calculated here are compared with experimental data in Table 5.1. As pointed out in Section 5.1 the calculated values are helpful in assessing the suitability of a membrane for a MPOD, even though a number of approximations and assumptions are made in obtaining them. The whole question of membrane permeability to gases is examined in more detail in a number of places, for example, references 2 and 3.

REFERENCES

1. D. W. van Krevelen, *Properties of Polymers*, Elsevier Publishing Company, Amsterdam, 1972.

2. J. Crank and G. S. Park, *Diffusion in Polymers*, Academic, New York, 1968.

3. H. B. Hopfenberg, Ed., *Permeability of Plastic Films and Coatings to Gases, Vapors and Liquids*, Plenum Press, New York, 1974.

VARIATION OF pH WITH THE TIME NEAR THE CATHODE OF A MPOD

We consider a detector of the type shown in Fig. 4.1a and make the approximations that the electrolyte film is a plane surface and that the cathode is a point source on this surface; these assumptions are reasonable if the radial distance from the center of the cathode through the electrolyte film to the electrolyte reservoir is much greater than both the film thickness and the cathode radius. We also assume the current has reached a steady value so that the cathode is providing a constant flux of OH^- ions. Making these assumptions we can now treat the diffusion of OH^- away from the cathode as diffusion from a continuous point source on an infinite plane surface.

Crank (1, Ch. 3) gives the equation for the concentration of a species at a distance r from an instantaneous point source on an infinite plane surface and for a time t after the deposition of the source

$$C_{r,t} = \frac{Q'}{4\pi Dt} \exp\left(-\frac{r^2}{4Dt}\right) \tag{J.1}$$

where Q' is the total amount of substance diffusing and D is the diffusion coefficient. By integrating this equation with respect to time we obtain $C_{r,t}$ corresponding to a continuous source

$$C_{r,t} = \frac{Q}{4\pi D} \int_0^t \exp\left(-\frac{r^2}{4Dt'}\right) \frac{dt'}{t'}$$

where Q is the rate of generation in mol s^{-1}. Letting $\tau' = r^2/4Dt'$ we obtain

$$C_\tau = \frac{Q}{4\pi D} \int_\tau^\infty \exp(-\tau') \frac{d\tau'}{\tau'} = \frac{Q}{4\pi D} E_1(\tau)$$

or

$$C_{r,t} = \frac{Q}{4\pi D} E_1\left(\frac{r^2}{4Dt}\right) \tag{J.2}$$

where $E_1(\tau)$ is the exponential integral (2, Ch. 5).

If we now assume a four-electron reduction of oxygen so that 10^5 C produce about 0.25 moles of OH^-, and if we also assume a current density of

3×10^{-5} A cm^{-2}, then with a cathode of radius, say, 0.1 cm, the rate of generation of OH$^-$ will be $Q \sim 2 \times 10^{-12}$ mol s^{-1}. Evaluating eq. (J.2) for $t = 10^3$ s (about 15 min) and $r = 0.1$ cm (the outer edge of the cathode) we have

$$C_{r,t} \simeq 2.0 \times 10^{-8} \text{ mol cm}^{-2}$$

Taking an electrolyte film thickness of ~ 10 μm, then the concentration of OH$^-$ at the outer edge of the cathode after running the detector for about 15 min in air-saturated water will be about 2×10^{-5} mol cm^{-3} or 20 mM. In an unbuffered solution this corresponds to pH = 12.3.

Thus we see that the local OH$^-$ concentration near the cathode of a MPOD rapidly reaches a high level in an unbuffered electrolyte. In a buffered electrolyte the rate of change of pH will obviously depend on the pH of the buffer, but for low pH buffers (e.g. <9) the local pH will still increase appreciably.

The influence of this pH change on the switch-on transient [Section 5.2.4] of a MPOD is shown in Fig. J.1. With a neutral solution (1M KCl) the first time the detector is switched on the attainment of a steady-state current is a very slow process, >10 min. After the detector has been running for 20 min, switched off for a minute or so, and then switched on again, the settling down time is much faster, being comparable to that obtained with a strongly alkaline electrolyte (0.5M KCl + 1M KOH—1:1). This effect is understandable in terms of changing diffusion conditions close to the electrode. If the electrolyte film thickness is having an effect on the size of the current being measured [Appendix H] then from eq. (H.6)

$$i_{L, a \neq 0} = \frac{i_L}{1 + K_0 \dfrac{D_m a}{D_e b}} \tag{H.6}$$

we can see that as D_e decreases and K_0 increases so the observed current, $i_{L, a=0}$, will decrease. As pH increases the solubility of oxygen falls due to salting-out [Section 2.4], and from eq. (4.3) we see that this will in fact cause K_0 to increase. Also the diffusion coefficient of oxygen does become smaller as pH rises (3). Therefore on this basis we would expect a continuous fall in the current as the pH increases until there is little further change in pH. There may of course be a counteracting effect arising from the fact that as the pH rises so the oxygen reduction will go more fully to completion, hydrogen peroxide will no longer be formed as an intermediate [Section 5.1.2], and n in eq. (4.17) will increase to a maximum value of four. However, since in fact the observed current is seen to fall steadily, it would seem that the dominating effect is that of changing diffusion conditions in the electrolyte at the cathode surface.

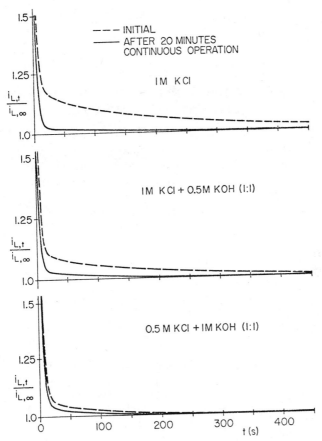

Fig. J.1 Switching-on transients for a MPOD with electrolytes of different pH.

REFERENCES

1. J. Crank, *The Mathematics of Diffusion*, Oxford University Press, 1956.
2. M. Abramowitz and I. A. Stegun, Eds., *Handbook of Mathematical Functions*, Dover Publications, New York, 1965.
3. R. E. Davis, G. L. Horvath, and C. W. Tobias, "The solubility and diffusion coefficient of oxygen in potassium hydroxide solutions," *Electrochim. Acta*, **12**, 287 (1967).

K

ELECTRODE POTENTIALS AND THERMODYNAMICS

For an electrode reaction the standard electrode potential, corresponding to equal concentrations of oxidant and reductant in the Nernst equation [eq. (3.9)], can be calculated on the basis of the thermodynamic data for the energy cycle, when the data are available (1, Ch. 2; 2). So, for example, for the case of metal dissolution and deposition we can draw the scheme

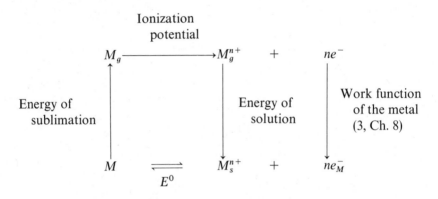

The subscripts g, s, and M denote gas, solution, and metal, respectively. It should be noted that in this scheme we have for convenience omitted the corresponding cycle for hydrogen; this should strictly be included because the standard electrode potential is referred to the H_2/H^+ couple.

Clearly, the electrochemical potential, E^0, of a reaction is related to the energy change, ΔG^0, associated with the reaction, and this relationship is expressed by

$$\Delta G^0 = -nFE^0 \qquad (K.1)$$

This equation, it should be noted, applies only when the electrode reaction is at equilibrium and not when a net current flows. Nevertheless, information about which direction the equilibrium will tend to move (i.e. oxidation of, say, a metal or reduction of its ion) can be obtained from the equation. As discussed in Appendix E an electrochemical reaction is no different in principle from a mechanical process in that spontaneous action occurs if the system can reduce its overall energy. So, just as a ball will roll down a slope to a position of minimum potential energy, a chemical reaction will proceed from left to right if the chemical energy of the products is less than that of the reactants. If ΔG^0 is expressed as $G_{prod} - G_{react}$, then clearly for a spontaneous process ΔG^0 is $-ve$, hence E^0 is $+ve$.

For the oxygen electrode reaction E^0 is positive,

$$O_2 + 2H_2O + 4e^- \rightleftharpoons 4OH^- \qquad E^0 = +0.401 \text{ V}$$

and eq. (K.1) shows that at an inert electrode in an oxygenated solution there will be a tendency for oxygen to be reduced and for there to be a flow of electrons through an external circuit; this is the thermodynamic basis of fuel cells using oxygen. On the other hand, for the $Cd/Cd(OH)_2$ couple,

$$Cd(OH)_2 + 2e^- \rightleftharpoons Cd + 2OH^- \qquad E^0 = -0.761 \text{ V}$$

E^0 is negative and so the tendency is for Cd metal to be oxidized and to form its hydroxide rather than reduction of the hydroxide to occur. Combining these two reactions which have opposite tendencies will provide a cell in which, on completion of an external circuit, oxygen will be reduced and cadmium will be oxidized,

$$O_2 + 2Cd + 2H_2O \rightleftharpoons 2Cd(OH)_2 \qquad E^0_{cell} = 1.162 \text{ V}$$

The potential for the overall reaction is obtained from the difference between the two standard potentials, a procedure that can be understood if one draws a vertical axis with the potentials above and below zero. E^0_{cell} is $+ve$, indicating the tendency of the reaction to proceed from left to right as written, and it is the value of the cell voltage that should theoretically be obtained if a measurement is made without drawing current from the cell; that is, under equilibrium conditions.

Finally, a word of caution. In the literature there are differences in conventions for the sign of the E^0 values. One convention is based on observed polarities, often called the European convention, while the other, the American convention, is based on energy considerations. However, provided charge transfer reactions are written as reductions,

$$M^{n+} + ne^- \rightleftharpoons M$$

the sign of the electrode potential as derived from an energy point of view agrees with that found from the observed polarity of the electrode. If this empirical rule is remembered no real problems with sign conventions should be met. Bockris and Reddy (3, Ch. 9) have given a useful and more detailed examination of the problems of signs of electrode potentials.

REFERENCES

1. B. E. Conway, *Theory and Principles of Electrode Processes*, Ronald, New York, 1965.
2. W. M. Latimer, *The Oxidation States of the Elements and their Potentials in Aqueous Solutions*, Prentice-Hall, New Jersey, 1961.
3. J. O'M. Bockris and A. K. N. Reddy, *Modern Electrochemistry*, Plenum Press, New York, 1970.

L

RESPONSE OF A MPOD TO A STEP CHANGE IN OXYGEN CONCENTRATION

The condition that we are considering is illustrated in Fig. 5.7, which is reproduced here.

Initially ($t < 0$) the oxygen concentration in solution is $C_{s, t=0}$, with the concentration within the membrane and at its outer face being $C_{m, t=0}$; $C_{s, t=0}$ and $C_{m, t=0}$ are related by the distribution coefficient K_b [Section 4.2]. A linear decrease of concentration with distance is assumed within the membrane and the inner surface of the membrane is approximated to the electrode surface [Section 5.2.4]. A step change in the oxygen concentration in solution occurs at $t = 0$, and provided that the equilibrium between oxygen in solution and oxygen in the membrane is maintained (i.e. the equilibrium is labile), then there is a step change in the surface concentration of the membrane to a new value, $C_{m, t>0}$. The concentration profiles within the membrane then vary with time until a linear concentration gradient over the total thickness of the membrane is once more attained.

This time-dependent situation can be described by the following differential equation and boundary conditions:

$$\frac{\partial C}{\partial t} = D \frac{\partial^2 C}{\partial x^2} \tag{L.1}$$

with

$$\frac{\partial C}{\partial x} = \frac{C_{m, t=0}}{b} \qquad \text{for } 0 \leqq x \leqq b \text{ and } t \to 0 \tag{L.2}$$

$$C = 0 \qquad \text{for } x = 0 \text{ and } t \geqq 0 \tag{L.3}$$

$$C = C_{m, t>0} \qquad \text{for } x = b \text{ and } t > 0 \tag{L.4}$$

The differential equation [eq. (L.1)] is the standard equation for time- and distant-dependent diffusion (1, Ch. 1). The boundary conditions [eqs. (L.2)–(L.4)] are obtained by inspection of Fig. 5.7. Equation (L.1) can be solved analytically (2) using, for example, the method of the separation of variables or by means of the Laplace transformation (1, Ch. 2). However, the problem of concentration variation within a membrane for a step concentration at one face of the membrane is not a new one, and Crank (1, Ch. 4) gives a

235

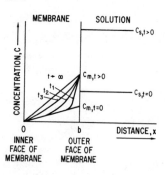

Fig. 5.7 Schematic diagram of concentration profiles within a membrane for a concentration step at one face of the membrane. Subscript "s" stands for "solution", and subscript "m" for "membrane".

general solution in the form of a trigonometrical series

$$C = C_{x=0} + (C_{m, t>0} - C_{x=0})\frac{x}{b}$$

$$+ \frac{2}{\pi} \sum_{n=1}^{\infty} \frac{C_{m, t>0} \cos(n\pi) - C_{x=0}}{n} \sin\left(\frac{n\pi x}{b}\right) \exp\left(-\frac{Dn^2\pi^2 t}{b^2}\right)$$

$$+ \frac{2}{b} \sum_{n=1}^{\infty} \sin\left(\frac{n\pi x}{b}\right) \exp\left(-\frac{Dn^2\pi^2 t}{b^2}\right) \int_0^b f(x') \sin\frac{n\pi x'}{b} dx' \qquad \text{(L.5)}$$

In our case $C_{x=0} = 0$ for all times and $f(x')$, which describes the initial variation of C with x in the region $0 < x < b$, is simply given by

$$C = \frac{C_{m, t=0} x'}{b} \qquad \text{(L.6)}$$

Substitution of eq. (L.6) into the integral of eq. (L.5) and the evaluation give (3, p. A181)

$$\int_0^b \frac{C_{m, t=0} x'}{b} \sin\left(\frac{n\pi x'}{b}\right) dx' = \frac{C_{m, t=0}}{b}\left[\frac{b^2}{n^2\pi^2} \sin\left(\frac{n\pi x'}{b}\right) - \frac{x'b}{n\pi} \cos\frac{n\pi x'}{b}\right]_0^b$$

$$= -\frac{C_{m, t=0} b^2 \cos(n\pi)}{bn\pi}$$

$$= -\frac{C_{m, t=0} b \cos(n\pi)}{n\pi} \qquad \text{(L.7)}$$

Now we are interested not so much in the variation of concentration with time and distance, but rather the variation of concentration gradient; that is, the current variation. Thus we need to substitute from eq. (L.7) into eq. (L.5) and to differentiate with respect to x:

$$\frac{\partial C}{\partial x} = \frac{C_{m, t>0}}{b} + \frac{2C_{m, t>0}}{b} \sum_{n=1}^{\infty} \cos(n\pi) \cos\left(\frac{n\pi x}{b}\right) \exp\left(-\frac{Dn^2\pi^2 t}{b^2}\right)$$

$$- \frac{2C_{m, t=0}}{b} \sum_{n=1}^{\infty} \cos(n\pi) \cos\left(\frac{n\pi x}{b}\right) \exp\left(-\frac{Dn^2\pi^2 t}{b^2}\right) \tag{L.8}$$

Subtracting $C_{m, t=0}/b$ from each side of eq. (L.8) and rearranging,

$$\frac{\dfrac{\partial C}{\partial x} - \dfrac{C_{m, t=0}}{b}}{\dfrac{C_{m, t>0}}{b} - \dfrac{C_{m, t=0}}{b}} = 1 + 2 \sum_{n=1}^{\infty} \cos(n\pi) \cos\left(\frac{n\pi x}{b}\right) \exp\left(-\frac{Dn^2\pi^2 t}{b^2}\right) \tag{L.9}$$

If we let $\tau = Dt/b^2$ and we replace $\cos(n\pi)$ by $(-1)^n$, since $\cos(n\pi)$ has a value of $+1$ or -1 according to whether n is even or odd, then we obtain eq. (5.7) from eq. (L.9):

$$\frac{\dfrac{\partial C}{\partial x} - \dfrac{C_{m, t=0}}{b}}{\dfrac{C_{m, t>0}}{b} - \dfrac{C_{m, t=0}}{b}} = 1 + 2 \sum_{n=1}^{\infty} (-1)^n \cos\left(\frac{n\pi x}{b}\right) \exp(-n^2\pi^2\tau) \tag{5.7}$$

REFERENCES

1. J. Crank, *The Mathematics of Diffusion*, Oxford University Press, 1956.
2. M. L. Hitchman, unpublished results.
3. *Handbook of Chemistry and Physics* (50th ed.), The Chemical Rubber Company, Cleveland, 1969.

PRINCIPLES OF CONDUCTANCE MEASUREMENTS

The relation between a current i (amperes) flowing in a metallic conductor and the potential or electromotive force, E (volts), between the ends of the conductor is given by Ohm's law

$$E = iR \qquad (M.1)$$

where the proportionality constant R is the resistance in ohms of the conductor. R depends both on the size of the conductor and on its composition, and this can be expressed by

$$R = \frac{\rho l}{A} \qquad (M.2)$$

In this equation l is the length and A the cross-sectional area of the conductor, and ρ is a constant. When l and A are both unity ρ is equal to the resistance of the unit conductor and depends only on the composition of the conducting material. This property ρ is called the *specific resistance* or *resistivity* of the conductor and it clearly has the units of ohm cm.

For liquids and solutions ρ may be defined in the same way as for a solid conductor, although it is more usual to use the reciprocal of ρ, the *specific conductance* or *conductivity*, $\varkappa$:

$$\varkappa = \frac{1}{\rho} \quad \text{and} \quad R = \frac{l}{A\varkappa} \qquad (M.3)$$

The units of $\varkappa$ are $\text{ohm}^{-1}\,\text{cm}^{-1}$. Some specific conductances are listed in Table M.1.

Measurements of the specific conductance of a liquid are made in a special cell which is inserted in one arm of a Wheatstone bridge—Fig. M.1. The bridge can be energized by a dc supply, in which case the dc resistance is measured by bringing the bridge into balance to give a null deflection on a galvanometer. It is more usual though for electrolyte solutions to use an ac supply with a frequency ~ 1 kHz in order to prevent the overall electrolysis that occurs with a direct current. Detection with an oscillator supply is visual with an oscilloscope or audio with headphones, and balance

TABLE M.1   Specific Conductance Data

a.   Variation with Material or Solution Composition

Material or solution	Specific conductance $\varkappa$ $(\text{ohm}^{-1}\,\text{cm}^{-1})$	Radius of hydrated cation (Å)
Ag	6.25×10^5	
Al	3.70×10^5	
Au	4.17×10^5	
Gas carbon	2×10^2	
Graphite	7.27×10^2	
Platinum	9.43×10^4	
LiCl $0.1M$; 25°C	9.59×10^{-3}	3.40
NaCl $0.1M$; 25°C	10.67×10^{-3}	2.76
KCl $0.1M$; 25°C	12.89×10^{-3}	2.32

b.   Variation with Temperature and Solution Concentration

g KCl/1000 g solution	$\varkappa$ $(\text{ohm}^{-1}\,\text{cm}^{-1})$		
	0°C	18°C	25°C
7.1135×10	6.517×10^{-2}	9.784×10^{-2}	1.113×10^{-1}
7.4191	7.138×10^{-3}	1.117×10^{-2}	1.286×10^{-2}
7.4526×10^{-1}	7.736×10^{-4}	1.221×10^{-3}	1.409×10^{-3}

is achieved when both the resistance and capacitance of the conductivity cell are correctly compensated. However, only the resistive component is used in calculating $\varkappa$. For liquids it is not easy to define l and A exactly, and this problem is overcome by measuring the resistance of a standard solution in the cell, which allows l/A, the cell constant, to be obtained. Further information on the theory and practice of conductivity measurements can be found in many texts (e.g. 1, Ch. I; 2, Ch. 5).

For an electrolyte solution the conductivity can readily be pictured as depending on both the number and speed, or mobility, of the ions—the more ions there are and the faster they move under an applied field the

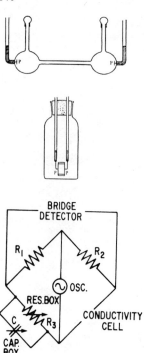

BRIDGE
DETECTOR

R_1 R_2

OSC.

RES.BOX

C R_3

CAP.
BOX

CONDUCTIVITY
CELL

Fig. M.1 Conductivity cells and a conductivity bridge.

higher will be the conductivity:

$$\varkappa \propto \sum_i C_i u_i \tag{M.4}$$

where u_i is the ionic mobility. Table M.1 gives values of the specific conductance for several electrolyte solutions of different concentrations and compositions. The linear dependence of $\varkappa$ on concentration is clearly seen, and when one compares the size of hydrated ions the proportionality to ionic mobility also becomes apparent—the larger the ion, the greater the viscous drag as it moves and the lower its mobility. Table M.1 also shows that the conductivity of electrolyte solutions is quite temperature dependent and that as the temperature increases so does the conductivity. This dependence arises from the fact that the mobility of an ion of radius r can be, to a first approximation, related to the viscosity of the medium in which it is moving by the Stokes equation,

$$u_i = \frac{1}{6\pi\eta r} \tag{M.5}$$

and the fact that the temperature dependence of the viscosity is given by

$$\eta = \eta_0 \exp\left(\frac{E_v}{RT}\right) \tag{F.3}$$

REFERENCES

1. D. R. Browning, Ed., *Electrometric Methods*, McGraw-Hill, London, 1969.
2. R. A. Robinson and R. H. Stokes, *Electrolyte Solutions*, Butterworths, London, 1970.

SYMBOLS

"One must accept that symbols will have to be used ambiguously sometimes, and hope that the reader will use his intelligence to place the appropriate meaning on them at the appropriate time."

Roger Fenn
"Symbolomania," *Nature*, 1970.

a	Coefficient in eq. (2.11)
a	Activity
a	Thickness of electrolyte film
a'	Coefficient in eq. (2.24)
a_2	Activity of solute
A	Coefficient in eq. (2.10)
A	Area
A'	Coefficient in power series for variation of heat of solution with temperature
b	Coefficient in eq. (2.11)
b	Thickness of membrane
b'	Coefficient in eq. (2.24)
B	Coefficient in eq. (2.10)
B'	Coefficient in power series for variation of heat of solution with temperature
c	Coefficient in eq. (2.11)
c'	Coefficient in eq. (2.24)
C	Concentration of dissolved oxygen
C	Coefficient in eq. (2.10)
C'	Coefficient in power series for variation of heat of solution with temperature
C_e	Concentration of dissolved oxygen in electrolyte solution
C_ε	Electrolyte concentration
C_i	Concentration of ionic species i
C_{in}	Dissolved oxygen concentration at input of coulometric cell
C_m	Concentration of oxygen within oxygen detector membrane
C_{out}	Dissolved oxygen concentration at output of coulometric cell
C_s	Concentration of dissolved oxygen in test solution

243

C_0	Concentration of electroactive species at surface of electrode
C_2	Concentration of solute
C_∞	Concentration of electroactive species far away from electrode surface
C_0^O	Concentration of oxidized species at electrode surface
C_0^R	Concentration of reduced species at electrode surface
C_∞^O	Concentration of oxidized species far away from electrode surface
C_∞^R	Concentration of reduced species far away from electrode surface
$C_{e, x=0}$	Dissolved oxygen concentration in electrolyte of MPOD at $x = 0$
$C_{m, t=0}$	Concentration of oxygen within membrane at time zero
$C_{m, t>0}$	Concentration of oxygen within membrane after step change in external oxygen concentration
$C_{m, x=b}$	Concentration of oxygen within membrane at $x = b$
$C_{m, x=0}$	Concentration of oxygen within membrane at $x = 0$
$C_{r,t}$	Concentration dependent on radial distance and time
$C_{s, t=0}$	Dissolved oxygen concentration in test solution at time zero
$C_{s, t>0}$	Dissolved oxygen concentration in test solution for all times other than zero
$C_{x,t}$	Concentration of electroactive species as a function of distance from electrode surface and time
d	Coefficient in eq. (2.11)
d	Depth of water
d'	Coefficient in eq. (2.24)
D	Diffusion coefficient
D_a	Diffusion coefficient of gas in amorphous polymer
D_e	Diffusion coefficient of oxygen in electrolyte film of MPOD
D_m	Diffusion coefficient of oxygen in membrane
D_0	Preexponential factor in eq. (3.23)
D_s	Diffusion coefficient of oxygen in test solution
D_{sc}	Diffusion coefficient of gas in semicrystalline polymer
e	Coefficient in eq. (2.11)
E	Electrode potential
E^0	Standard electrode potential
E_{app}	Applied potential
E_{cell}	Cell potential
E_D	Activation energy for diffusion
E_e	Equilibrium electrode potential
E_P	Activation energy for gas permeation in membrane
E_v	Activation energy for viscous flow

E_P'	Activation energy for combined processes of gas permeation in membrane and gas dissolution in solution
f	Fugacity
f	Flow rate
f_i	Fugacity of species i
f_2	Fugacity of solute
f_2^0	Fugacity of solute in the standard state
F	Faraday
F	Force
g	Gravitational constant
ΔG^0	Standard free energy change
h	Height above a reference level
ΔH	Heat of solution
ΔH_g	Heat of solution of gas in a glassy polymer
ΔH_m	Heat of solution of gas in membrane
ΔH_r	Heat of solution of gas in elastomer
ΔH_v	Heat of vaporization
i	Current
i_a	Anodic current
i_c	Cathodic current
i_L	Transport limited current
i_0	Exchange current
i_t	Time-dependent current
i_0^0	Standard exchange current
$i_{L, a \neq 0}$	Limiting current at MPOD with nonzero electrolyte thickness
$i_{L,in}$	Limiting current at input to coulometric cell
$i_{L,0}$	Limiting current under initial conditions
$i_{L,t}$	Time-dependent limiting current
$i_{L,\theta}$	Limiting current at temperature $\theta°C$
$i_{L,\infty}$	Limiting current at long times
I	Ionic strength
J	Diffusional flux
J_e	Diffusional flux through electrolyte layer of MPOD
J_m	Diffusional flux through membrane
k	Proportionality constant
k	Henry's law proportionality constant
k'	Proportionality constant in approximate form of Henry's law
k^0	Heterogeneous rate constant at standard electrode potential
k_b	Heterogeneous rate constant for reverse reaction
k_f	Heterogeneous rate constant for forward reaction
k_s	Salting coefficient
k_b^0	Standard heterogeneous rate constant for reverse reaction

k_f^0	Standard heterogeneous rate constant for forward reaction
K	Proportionality constant
K_b	Distribution coefficient for oxygen at test solution/MPOD membrane interface
K_0	Preexponential factor in eq. (5.14)
K_0	Distribution coefficient for oxygen at electrolyte layer—membrane interface
K_0'	Preexponential factor in eq. (5.16)
l	Characteristic length
m	Flow rate of mercury at dropping mercury electrode
M	Molecular or atomic weight
M_A	Molecular weight of air
M_2	Molecular weight of solute
MPOD	Membrane polarographic oxygen detector
n	Number of electrons transferred in electrochemical reaction
N_i	Number of moles of component i
N_O	Number of moles of oxidized species
N_R	Number of moles of reduced species
O	Oxidized species
p	Partial or vapor pressure
p	Coefficient in eq. (2.23)
p'	Coefficient in eq. (2.24)
p_i	Partial or vapor pressure of component i
p_1	Vapor pressure of solvent
p_2	Vapor pressure of solute
p_2^0	Reference vapor pressure of solute
P	Total gas pressure
P^0	Reference pressure
P_h	Atmospheric pressure at height h
P_m	Permeability coefficient of oxygen in membrane
P_0	Preexponential factor in eq. (5.15)
P_r	Atmospheric pressure at reference height
P_T	Barometric pressure
P_0'	Preexponential factor in eq. (5.17)
Pe	Peclet number
Pr	Prandtl number
q	Coefficient in eq. (2.23)
q	Charge
q'	Coefficient in eq. (2.24)
Q	Rate of generation of OH^- ions
Q'	Amount of substance diffusing
r	Coefficient in eq. (2.23)

r	Radius
r	Radial distance
r'	Coefficient in eq. (2.24)
R	Molar gas constant
R	Reduced species
R_{ct}	Charge transfer resistance
R_0	Preexponential factor in eq. (5.20)
R_T	Thermistor resistance at thermodynamic temperature T
Re	Reynolds' number
s	Coefficient in eq. (2.23)
s	Sensitivity coefficient for coulometric cell
s'	Coefficient in eq. (2.24)
s_T	Sensitivity coefficient of coulometric cell at thermodynamic temperature T
S	Solubility of oxygen (mg liter^{-1}) in air-saturated water at total atmospheric pressure of 760 torr
S'	Solubility of oxygen (mg liter^{-1}) in air-saturated water at total pressure other than 760 torr
S_e	Oxygen solubility in salt solution
S_0	Preexponential factor in eq. (I.6)
S_1	Oxygen solubility for oxygen partial pressure of 760 torr
S_2	Oxygen solubility for oxygen partial pressure of $(760 - p_1)$ torr; that is, corrected for water vapor pressure
$S_{a,e}$	Gas solubility in amorphous elastomer
$S_{a,g}$	Gas solubility in glassy, amorphous polymer
$S_{sc,g}$	Gas solubility in semicrystalline, glassy polymer
Sc	Schmidt number
t	Coefficient in eq. (2.23)
t	Time
t_0	Time zero
t_∞	Long time
t_{wt}	Wear-out time for auxiliary electrode of MPOD
T	Thermodynamic temperature
T'	Thermodynamic temperature of state represented by $'$
T''	Thermodynamic temperature of state represented by $''$
T_b	Boiling point
T_g	Glass transition temperature of polymer
T_r	Temperature at reference height
T_{cr}	Critical temperature of gas
u_i	Mobility of ion i
v_c	Critical velocity for change from laminar to turbulent flow
v_m	Main stream velocity

V	Volume
V_c	Critical voltage
w	Amount of material converted electrochemically
x	Length
x_c	Degree of crystallinity of polymer
x_i	Mole fraction of component i
x_2	Mole fraction of solute
x_2^0	Reference mole fraction of solute
x_2'	Mole fraction of solute in state represented by '
x_2''	Mole fraction of solute in state represented by "
$\bar{x}_2$	Mole fraction of solute in vapor over solution
z_i	Charge of ion i
α	Bunsen absorption coefficient
α	Transfer coefficient
β	D/δ for coulometric cell
β'	$DA/V\delta$ for coulometric cell
γ	Activity coefficient for gas in pure water
γ_c	Activity coefficient based on concentration as measure of composition
γ_e	Activity coefficient of gas in electrolyte solution
γ_p	Activity coefficient based on partial pressure as measure of composition
γ_x	Activity coefficient based on mole fraction as measure of composition
δ	Diffusion boundary layer thickness
δ_0	Hydrodynamic boundary layer thickness
δ_t	Thickness of time-dependent diffusion layer
δ_T	Variation of temperature with height
ε	Exponent in eq. (3.35)
ζ	Exponent in eq. (3.35)
η	Overpotential
η	Coulombic efficiency
η	Coefficient of viscosity
η_0	Preexponential factor in eq. (F.3)
θ	Temperature in °C
$\varkappa$	Specific conductance
λ	Coefficient in eq. (5.25)
μ	Chemical potential
μ^0	Chemical potential at unit activity
μ^*	Chemical potential at unit fugacity
μ_i	Chemical potential of species i
ν	Kinematic viscosity

$\bar{v}_2$	Partial molal volume of gas
$\bar{v}_2^\infty$	Partial molal volume of gas at reference composition of infinite dilution
ξ	Coefficient in eq. (5.25)
ρ	Density
ρ	Specific resistance
τ	Time variable
φ	Sensitivity coefficient for MPOD
φ_T	Sensitivity coefficient at thermodynamic temperature T
φ_θ	Sensitivity coefficient at temperature $\theta°C$
χ	Coefficient in eq. (5.25)
ψ	Integration variable
ω	Coefficient in eq. (5.25)

INDEX